Some Observations On Risk-Informing Appendices A & B to 10 CFR Part 50

Prepared for the Advisory Committee on Reactor Safeguards

U.S. Nuclear Regulatory Commission
Advisory Committee on Reactor Safeguards
Washington, DC 20555-0001

Some Observations On Risk-Informing Appendices A and B to 10 CFR Part 50

Prepared for the Advisory Committee on Reactor Safeguards

Manuscript Completed: December 2001
Date Published: January 2002

Prepared by:
J.N. Sorensen
Senior Fellow

Advisory Committee on Reactor Safeguards
U.S. Nuclear Regulatory Commission
Washington, DC 20555-0001

ABSTRACT

This report was prepared for the Advisory Committee on Reactor Safeguards to provide a basis for discussing possible changes to the general design criteria (GDC) of Appendix A to Title 10, Part 50, of the *Code of Federal Regulations* in order to make them more consistent with a risk-informed regulatory structure. The three broad options identified include (1) changing the scope of the GDC from "important to safety" to "important to risk," (2) modifying individual criteria to address risk more directly, and (3) replacing the GDC with safety goals and risk acceptance criteria. As written, the GDC apply to systems, structures, and components (SSCs) important to safety, which are defined as those SSCs that are required to provide reasonable assurance that there will be no undue risk to public health and safety. If the scope is changed to "important to risk," the GDC can be applied to those SSCs that will have the greatest impact on reducing risk, as measured by risk metrics such as core damage frequency or large early release frequency. Specific changes to improve the risk focus of individual criteria are also discussed. Such changes might involve replacing requirements for redundancy, diversity, and independence with an overall reliability goal for a system or function. Applicability of the GDC to non-light-water reactors (LWRs) is briefly examined, with the conclusion that slightly more than half of the criteria could apply to non-LWRs, but the remainder should be modified or replaced to address phenomena important to the safety of other reactor types. Another brief discussion of the quality assurance requirements of Appendix B to 10 CFR Part 50 concludes that Appendix B has sufficient flexibility to permit less stringent requirements for SSCs that are less important to risk.

CONTENTS

Attachments

ABBREVIATIONS

ACRS	Advisory Committee on Reactor Safeguards
ASME	American Society of Mechanical Engineers
ATWS	anticipated transient without scram
BI	barrier integrity
BWR	boiling-water reactor
CD	core damage
CDF	core damage frequency
CFR	*Code of Federal Regulations*
CPR	control of power or reactivity
CRGR	Committee to Review Generic Requirements
DiD	defense-in-depth
ECCS	emergency core cooling system
EP	emergency planning
FPC	fission product containment
GDC	general design criteria
HR	heat removal
IE	initiating event
IEEE	Institute of Electrical and Electronics Engineers
LERF	large early release frequency
LOCA	loss-of-coolant accident
LR	large release
LWR	light-water reactor
MS	mitigating system
MTS	marginal to safety
NEI	Nuclear Energy Institute
NRC	U.S. Nuclear Regulatory Commission
PRA	probabilistic risk assessment
PWR	pressurized-water reactor
QA	quality assurance
QAC	quality assurance criteria
RCI	reactor coolant inventory
RCPB	reactor coolant pressure boundary
RISC	risk-informed safety class
RRG	Regulatory Review Group
SFC	single failure criterion
SR	small release
SRM	staff requirements memorandum
SSC	structure, system, and component

1. INTRODUCTION

Over the past several years, the U.S. Nuclear Regulatory Commission (NRC) has been working to develop a regulatory structure and supporting processes that will make greater use of risk information than is presently the case. Like earlier regulatory reform efforts, the current work on developing risk-informed regulations recognizes the central role that the first two appendices to Title 10, Part 50, of the *Code of Federal Regulations* (10 CFR Part 50) play in the regulatory structure and the importance of ensuring that any changes contemplated for Part 50 are properly supported by changes to the appendices. This report examines Appendix A, "General Design Criteria for Nuclear Power Plants," and Appendix B, "Quality Assurance Criteria for Nuclear Power Plants and Fuel Reprocessing Plants," to determine the degree to which they support risk-informed regulation, and what changes might be necessary or desirable to provide better support. Recently, nuclear industry interest in the pebble-bed modular reactor concept has raised separate questions regarding the applicability of the current regulatory structure to reactors other than light-water moderated and cooled designs. A brief discussion of that issue is also provided.

At the risk of stating the obvious, the best way to risk-inform Appendices A and B will depend on what other changes are made to the body of Part 50 to make it more risk-informed. The NRC staff's current work to risk-inform both the special treatment requirements in Part 50 (Option 2 of SECY-98-300) [1][1] and the technical requirements in Part 50 (Option 3) is expected to result in recommended changes to Appendix A and possibly Appendix B. In order to establish reasonable boundaries on the scope of this report, however, both Appendices A and B are discussed without considering supporting or conforming changes in the rest of Part 50.

Section 2 examines the scope, content, and structure of Appendix A to develop insights regarding the relationship between the general design criteria (GDC) as they are currently written and risk-informed regulation. In particular, Section 2 examines how the GDC relate to currently contemplated risk metrics, such as core damage frequency (CDF) and large early release frequency (LERF). Section 3 provides a brief summary of earlier regulatory reform efforts, and how Appendices A and B were viewed in the context of suggested regulatory changes. Section 4 provides a brief summary of current NRC staff initiatives to develop alternatives for making risk-informed changes to 10 CFR Part 50, and how that work might touch on the GDC and the quality assurance criteria (QAC) of Appendices A and B. Section 5 identifies three broad options for revising the GDC to better support risk-informed regulation. Section 6 discusses the individual design criteria, whether each poses an impediment to risk-informed regulation, and how each might be changed to improve its focus on risk metrics. Section 7 provides overall conclusions regarding risk informing the GDC. Sections 8 and 9 deal briefly with the applicability of the GDC to non-light-water moderated and cooled reactors, and the need for changes to Appendix B to support risk-informed regulation, respectively.

---

[1]Numbers in square brackets identify references listed in Section 10 of this report.

2. DESCRIPTION OF THE GENERAL DESIGN CRITERIA

The GDC comprise 55 individual criteria, which are grouped into the following 6 categories:

I   Overall Requirements
II   Protection by Multiple Fission Product Barriers
III   Protection and Reactivity Control Systems
IV   Fluid Systems
V   Reactor Containment
VI   Fuel and Radioactivity Control

There is some overlap between the categories, and some criteria don't fit neatly within their assigned category. There is also some variation among categories with regard to how the criteria are structured. For example, a single criterion covers the inspection and testing of electrical power systems. Emergency core cooling systems, however, are accorded a criterion for inspection and a separate criterion for testing. Of the 55 criteria, 13 address inspection or testing of structures, systems, or components (SSCs) which are themselves required by other criteria. For ease of reference, Attachment 1 to this report contains the complete text of Appendix A to 10 CFR Part 50, "General Design Criteria for Nuclear Power Plants."

The GDC were initially incorporated into Part 50 in February 1971, although draft versions were publicly available as early as 1965. A 1965 draft, provided to the ACRS and the Commission for consideration, contained 27 criteria [2]. (Attachment 2 to this report contains the text of the 1965 draft criteria.) A 1969 draft expanded the number of criteria to 70. Later revisions combined some criteria and deleted others, but attempted to preserve something of the structure and numbering scheme for the 70 criteria in the earlier draft. Among other things, this resulted in gaps in the numbering system; there are no criteria numbered between 6 and 9, for example.

As required by 10 CFR 50.34, applications for construction permits for nuclear power plants must include the principal design criteria for the facility. As stated in Appendix A, the GDC establish minimum requirements for the principal design criteria for water-cooled nuclear power plants. Appendix A also states that (1) the GDC are considered to be "generally applicable" to other types of nuclear power units, (2) the development of the GDC is not yet complete, (3) there will be some water-cooled nuclear power plants for which the GDC are not sufficient, and (4) there may be water-cooled nuclear plants for which fulfillment of some of the GDC may not be necessary. The issues identified in Appendix A as requiring additional consideration are single failures of passive components; redundancy and diversity requirements for fluid systems important to safety; type, size, and orientation of possible breaks in the reactor coolant pressure boundary (RCPB); and the possibility of systematic, non-random, concurrent failures of redundant elements in control systems.

The broad scope of the GDC is "structures, systems, and components important to safety," which are then defined as "structures, systems, and components that provide reasonable assurance that the facility can be operated without undue risk to the health and safety of the public." The standard of "no undue risk" is equivalent to the "adequate protection" standard derived from the Atomic Energy Act. Compliance with the body of the Commission's regulations is presumed to result in adequate protection. Thus, the performance standard addressed by the GDC is adequate protection, rather than the Commission's safety goals or other risk metrics.

The clear intent of the GDC is to reduce the probability and consequences of accidents that could affect public health and safety. The means for doing so are qualitative rather than quantitative, and reflect the state-of-the-art of reactor design in the mid-1960s. Rather than establishing quantitative requirements for the reliability of safety systems, the GDC require that redundancy and diversity be provided for safety-significant functions, with the presumed result being improved reliability. Embedded in the GDC is what is often referred to as the "single failure criterion" (SFC). Nine of the criteria specify that a system must be capable of performing its safety function "assuming a single failure." Appendix A defines "single failure" as "an occurrence which results in the loss of capability of a component to perform its intended safety functions." Multiple failures resulting from a single occurrence are considered to be a single failure.

A reasonable first step in trying to determine the relationship between the GDC and risk-informing Part 50 is to examine the structure and content of the criteria, and see how they relate to current risk-informed terminology. The GDC embody four different kinds of information. The first is to establish the importance of conservative, stable, forgiving designs. The second is to provide a catalog of conditions that the design must accommodate. The third is to establish redundancy and diversity requirements for important safety functions, with the implicit assumption that appropriate redundancy and diversity will result in higher system reliability. The fourth is to establish defense-in-depth requirements, both in the sense of multiple fission product barriers and in the sense of providing several systematic layers of defense (preventing accident initiators, terminating accident sequences, and mitigating accident consequences). These attributes can be summarized as conservatism, capability, reliability, and defense-in-depth. Table 1 provides a list of the GDC, and the first data column classifies each criterion according to these four attributes.

Another way of classifying the GDC is to look at what is required or specified by each criterion. With a few exceptions, each criterion requires one of three things: (1) a function must be provided; (2) margin or stability must be provided; or (3) certain systems must be inspected, certain functions must be tested, or provisions must be made for inspection or testing. Columns 2, 3, and 4 of Table 1 classify each of the GDC according to this perspective.

The GDC can also be classified in terms of the safety function or safety goal addressed, or (in the terminology of the Reactor Oversight Program) the reactor safety cornerstone addressed. Column 5 of Table 1 associates each criterion with a safety function, namely control of power or reactivity (CPR), heat removal (HR), fission product containment (FPC), control of reactor

3

TABLE 1

Characteristics of General Design Criteria

Column Number/Notes on Sheet 2	Attribute (1)	Requires Function (2)	Requires Margin (3)	Req. Test (4)	Safety Function (5)	Safety Goal (6)	Corner-stone (7)	Modify Scope (8)	Recast in Risk Terms? (9)	Apply to Non-LWRs? (10)
I Overall Requirements										
1 Quality Standards and Records	Reliability				all	all	all	1	Scope	yes
2 Design Bases for Protection Against Natural Phenomena	Capability				all	all	all	1	Scope	yes
3 Fire Protection	Capability	yes			all	all	all	1	Scope	yes
4 Environmental and Dynamic Effects Design Bases	Capability				all	all	all	1,2	Scope	yes
5 Sharing of Structures, Systems, and Components	Reliability				all	all	all	1	Recast	yes
II Protection by Multiple Fission Product Barriers										
10 Reactor Design	Conservatism		margin		CPR	CD	IE		No change	yes
11 Reactor Inherent Protection	Conservatism		stability		CPR	CD	IE		No change	yes
12 Suppression of Reactor Power Oscillations	Capability	yes	stability		CPR	CD	IE		No change	yes
13 Instrumentation and Control	Capability				all	all	IE	5	Scope	yes, but
14 Reactor Coolant Pressure Boundary	Conservatism		margin	yes	HR	CD	IE, BI		No change	yes
15 Reactor Coolant System Design	Capability		margin		HR	CD	IE, BI		No change	yes
16 Containment Design	Capability, DiD	yes			FPC	LR	MS, BI	1	Recast	maybe
17 Electric Power Systems	Reliability	yes			all	all	IE, MS	1,3	Recast (SFC)	yes
18 Inspection and Testing of Electric Power Systems	Reliability			yes	all	all	IE, MS	1	Recast	yes
19 Control Room	Capability	yes		yes	all	LR	MS, BI	2	Recast	yes, but
III Protection and Reactivity Control Systems										
20 Protection System Functions	Capability	yes			CPR	CD	IE	1	No change	yes
21 Protection System Reliability and Testability	Reliability			yes	CPR	CD	IE		Recast (SFC)	yes
22 Protection System Independence	Reliability		margin		CPR	CD	IE		No change	yes
23 Protection System Failure Modes	Reliability		margin		CPR	CD	IE		No change	yes
24 Separation of Protection and Control Systems	Reliability				CPR	CD	IE		Recast (SFC)	yes
25 Protection System Requirements for Reactivity Control Malfunctions	Conservatism	yes			CPR	CD	IE		Recast	yes, but
26 Reactivity Control System Redundancy and Capability	Capability				CPR	CD	IE		Recast	no
27 Combined Reactivity Control Systems Capability	Capability		margin		CPR	CD	IE		No change	no
28 Reactivity Limits	Conservatism		margin		CPR	CD	IE		No change	yes, but
29 Protection Against Anticipated Operational Occurrences	Conservatism		margin		CPR	CD	IE		No change	yes
IV Fluid Systems										
30 Quality of Reactor Coolant Pressure Boundary	Reliability			yes	RCI	CD	BI	4	No change	yes, but
31 Fracture Prevention of Reactor Coolant Pressure Boundary	Capability		margin		RCI	CD	BI		No change	yes, not
32 Inspection of Reactor Coolant Pressure Boundary	Reliability			yes	RCI	CD	BI		No change	yes, but
33 Reactor Coolant Makeup	Capability	yes			RCI	CD	IE		Recast(?)	yes, but
34 Residual Heat Removal	Capability	yes			HR	CD	IE		Invokes SFC	yes, but
35 Emergency Core Cooling	Capability, DiD	yes			HR	CD	MS	2	Invokes SFC	no
36 Inspection of Emergency Core Cooling System	Reliability			yes	HR	CD	MS		No change	no
37 Testing of Emergency Core Cooling System	Reliability			yes	HR	CD	MS		No change	no
38 Containment Heat Removal	Capability, DiD	yes			FPC	LR	MS, BI	2	Recast (SFC)	no
39 Inspection of Containment Heat Removal System	Reliability			yes	HR	LR	MS, BI		No change	no
40 Testing of Containment Heat Removal System	Reliability			yes	HR	LR	MS, BI		Recast(?)	no
41 Containment Atmosphere Cleanup	Capability, DiD				FPC	LR	MS		Invokes SFC	no
42 Inspection of Containment Atmosphere Cleanup Systems	Reliability			yes	FPC	LR	MS		Recast (SFC)	no
43 Testing of Containment Atmosphere Cleanup Systems	Reliability			yes	FPC	LR	MS		No change	no
44 Cooling Water	Capability	yes			HR	CD	MS	1	Invokes SFC	yes

Sheet 1 of 2

4

TABLE 1

Characteristics of General Design Criteria

Column Number/Notes on Sheet 2	Attribute (1)	Requires Function (2)	Requires Margin (3)	Req. Test (4)	Safety Function (5)	Safety Goal (6)	Corner-stone (7)	Modify Scope (8)	Recast in Risk Terms? (9)	Apply to Non-LWRs? (10)
45 Inspection of Cooling Water System	Reliability			yes	HR	CD	MS		No change	yes
46 Testing of Cooling Water System	Reliability			yes	HR	CD	MS	2	No change	yes
V Reactor Containment										
50 Containment Design Basis	Capability, DiD	yes			FPC	LR	BI	2	No change	no
51 Fracture Prevention of Containment Pressure Boundary	Reliability		margin		FPC	LR	BI		No change	no
52 Capability for Containment Leakage Rate Testing	Reliability			yes	FPC	LR	BI		Recast(?)	no
53 Provisions for Containment Testing and Inspection	Reliability			yes	FPC	LR	BI		No change	no
54 Systems Penetrating Containment	Reliability	yes		yes	FPC	LR	BI	1	Recast	no
55 Reactor Coolant Pressure Boundary Penetrating Containment	Reliability	yes			FPC	LR	BI		Recast	no
56 Primary Containment Isolation	Reliability	yes			FPC	LR	BI		Recast	no
57 Closed Systems Isolation Valves	Reliability	yes			FPC	LR	BI		Recast	no
VI Fuel and Radioactivity Control										
60 Control of Releases of Radioactive Materials to the Environment	Capability	yes			FPC	SR	BI		No change	yes
61 Fuel Storage and Handling and Radioactivity Control	Capability	yes		yes	FPC	SR	BI	1	No change	yes
62 Prevention of Criticality in Fuel Storage and Handling	Capability	yes			FPC	SR	BI		No change	yes
63 Monitoring Fuel and Waste Storage	Capability	yes			FPC	SR	BI		No change	yes
64 Monitoring Radioactivity Releases	Capability	yes			FPC	LR, SR	BI, EP		No change	yes

Notes

(1) The design attribute(s) addressed by each criterion. DiD denotes defense-in-depth.

(2) Requires provision of a system or function.

(3) Requires design conservatism (stability or margin).

(4) Requires inspection, testing, or provision of capability to do so.

(5) The primary safety function addressed by each criterion: heat removal (HR), control of power or reactivity (CPR), fission product containment (FPC), control of reactor coolant inventory (RCI), or all of the above.

(6) The primary focus of each criterion: core damage (CD), large releases (LR), small releases (SR), or all of the above.

(7) Associates each criterion with the reactor safety cornerstones: initiating events (IE), barrier integrity (BI), mitigating systems (MS), emergency planning (EP) or all of the above

(8) 1 - Scope is "important to safety" or addresses "adequate safety."
2 - References LOCA, defined as including double-ended break of largest pipe.
3 - Requires electric power to be available "within a few seconds" following LOCA.
4 - Could be modified to require testing commensurate with importance to risk.
5 - Stated goal is to "assure adequate safety."

(9) Identifies whether a criterion can be usefully cast in risk terms.
"Scope" means that the scope could be changed from "important to safety" to "important to risk."
"Invokes SFC" means that the criterion requires safety function success assuming a single failure.
"Recast (SFC)" means that the criterion requires safety function success assuming a single failure, and should also be recast for other reasons.

(10) Identifies whether a criterion applies (unambiguously) to non-LWRs.

coolant inventory (RCI), or all safety functions. Similarly, Column 6 associates each criterion with a safety goal, namely core damage (CD), large release (LR), small release (SR), or all of the above. All of the criteria have a fairly obvious relationship to preventing core damage, preventing the uncontrolled release of radioactivity, or both. This perspective, particularly as it relates to the current articulation of large early release frequency (LERF) as a subsidiary safety goal, turns out to be a little strained. The relationship of 26 criteria to reducing CDF seems clear enough. The 17 criteria intended to reduce the probability of large releases, however, don't differentiate between early or late releases. Furthermore, the five criteria in the "Fuel and Radioactivity Control" group primarily address releases that one would expect to be small. Their apparent focus is control of chronic worker and public exposure, rather than risk.

Classifying the GDC with respect to the cornerstones of the Reactor Oversight Process turns out to work reasonably well. The results are shown in Column 7, where initiating events are denoted by IE, barrier integrity by BI, mitigating systems by MS, and emergency planning by EP. An attempt to go one step lower in the hierarchy by associating each criterion with one or more of the reactor oversight process performance indicators, was not very successful. For the most part, the GDC do not reflect the performance indicators because the performance indicators reflect operations and maintenance, rather than design.

Presuming that interest in risk-informing the GDC arises from the desire to improve the coherence of the regulatory structure, improve safety, or reduce regulatory burden without significantly reducing safety, it may be useful to consider the GDC in the context of the entire body of regulatory practice. While the GDC are important, they represent only a small portion of the design requirements embodied in the regulatory structure. Part 50, for example, incorporates the ASME code by reference, as well as certain standards promulgated by the Institute of Electrical and Electronics Engineers. Also, NRC has published about 160 Division 1 regulatory guides, 140 of which address issues related to one or more of the GDC. These regulatory guides, in turn, reference a number of industry standards. Risk-informing, or in the extreme even eliminating, the GDC, would quite likely to have the same kind of limited impact that the South Texas Project achieved in developing a graded quality assurance (QA) program. It might also be observed that there is a degree of uncertainty in our knowledge of how the current population of plants complies with the GDC. In fact, the Committee to Review Generic Requirements (CRGR) has noted that more than half of the plants that are currently operating received their construction permits before the GDC were added to Part 50.

3. PRIOR REGULATORY REFORM EFFORTS

Before discussing the GDC in detail, it may be worthwhile to summarize earlier regulatory reform efforts that touched on Appendices A and B. Both the GDC and the QAC have been considered important elements in earlier explorations of regulatory reform. For example, the stated goal of NRC's Program for Elimination of Requirements Marginal to Safety (MTS), initiated in 1984, was "Existing regulatory requirements that have marginal importance to safety should be eliminated" [3]. Specifically, that program identified and analyzed the risk significance of the following seven issues:

(1)   containment leak-rate testing
(2)   fire protection
(3)   equipment qualification
(4)   combustible gas control
(5)   quality assurance
(6)   physical protection
(7)   requests for information under 10 CFR 50.54(f)

The first four of these issues are each addressed by one or more of the GDC. The fifth issue, quality assurance, is addressed by GDC-1 and the entirety of Appendix B.

In 1991, the NRC staff recommended that the Commission close the MTS program, concluding that "... no 10 CFR Part 50 regulations were identified that are so burdensome on operating reactors and so marginal to safety that (they) would warrant the expenditure of staff resources to rectify" [4]. The staff later changed its conclusion as a result of public comments, and recommended an ongoing program to eliminate requirements marginal to safety [5]. Under that program, the staff recommended completing three rulemakings in the first year cycle, including containment leak rate testing, fire protection, and combustible gas control. A 1993 workshop on requirements marginal to safety showed that there was also substantial industry support for changing QA requirements. Ultimately, only containment leak rate testing (Appendix J) was revised as a direct result of the MTS program, and the remaining issues were included in a plan to institutionalize regulatory improvement [6]. Quality assurance was later addressed in a joint NRC/industry effort on graded QA, and the South Texas Project developed a graded QA program. No changes were made to Appendix B as a result of these efforts because it was concluded that Appendix B, as written, permitted graded QA programs. The industry and the NRC staff have never really come to closure on the need for changes to Appendix B.

Beginning in January 1993 and extending through February 1996, the Regulatory Review Group (RRG) undertook a broad scope effort to identify candidate requirements for regulatory reform [7]. The RRG examined the regulations in 10 CFR Parts 21, 26, 50, and 73, as well as the Division 1 Regulatory Guides and representative operating reactor licenses. The RRG also looked at the integration of risk analysis techniques into the regulatory process. The primary focus of this RRG work was to identify areas where regulatory burden could be reduced in a manner that was "safety neutral," and to look for opportunities to make the regulations more performance-based. The RRG concluded that "the rules themselves were not a major source of unnecessary burden on operating reactors." In particular, the RRG concluded that Appendices A and B were performance-based, substantially contributed to safety, and did not go beyond what was required for safety. The RRG further concluded that a major source of burden was staff requirements or licensee commitments that went beyond what the regulations actually required. This burden was aggravated by the staff's practice of enforcing over-commitments, once they were made. The RRG's recommended solution hinged on adopting (or recognizing) the following three principles:

(1)   The regulations themselves represent the safety standard to which licensees should be held accountable.

(2)   The licensee retains primary responsibility for compliance with the safety standard established by the regulations.

(3)   The amount of regulatory oversight should bear a relationship to the safety significance of the regulation.

The RRG basically recommended shifting the regulatory paradigm from one of pre-approval of all changes in regulatory programs to one of post-implementation review. They further recommended changing Part 50 to include the definition of "commitment" and establishing a change process for commitments. The basic idea was that licensees should be able to change their facilities and programs as long as they comply with the underlying regulations. By contrast, for example, licensees cannot change their QA programs without NRC approval if the change would reduce the commitments in the program. The standard against which change is measured is the currently approved program for a particular licensee, rather than the regulations themselves. The Commission never acted upon the RRG recommendations in this area.

About 3 years ago, the Nuclear Energy Institute (NEI) looked at the issue of making 10 CFR Part 50, including Appendices A and B, risk-informed and performance-based. The NEI's proposed approach involved redefining key phrases, such as "important to safety," "adequate safety," and "important features" in terms of risk. For example, NEI suggested defining an SSC important to safety to be consistent with "a structure, system or component which operating experience or probabilistic risk assessment has shown to be significant to public health and safety."

4. CURRENT STAFF ACTIVITIES

Current NRC initiatives to risk-inform the regulatory process were proposed to the Commission in SECY-98-300 [1]. In the staff requirements memorandum (SRM) for that Commission paper [8], the Commission directed the staff to proceed with implementing three options related to risk-informed revisions to 10 CFR Part 50. The first option was to continue certain ongoing rulemaking activities, rather than delaying them in anticipation of possible additional changes on the basis of risk information. The then ongoing rulemakings, which were to proceed without consideration of additional risk information, were 10 CFR 50.59 (Changes, tests, and experiments), 10 CFR 50.72 (Immediate notification requirements for operating nuclear power reactors), 10 CFR 50.73 (Licensee event report system), 10 CFR 50.55a (Codes and standards), and 10 CFR 50.67 (Accident source term). The second option was to identify changes to the scope of SSCs covered by those sections of Part 50 requiring special treatment such as QA, environmental qualification, technical specifications, or the Boiler and Pressure Vessel Code promulgated by the American Society of Mechanical Engineers (ASME). Specifically, SSCs that were determined to be of low safety significance would no longer be subject to special treatment requirements, but would be treated as commercial grade. The third option was to consider including risk-informed attributes in specific regulatory requirements in Part 50. This option could be as broad as a complete rewrite of Part 50, or it could focus on regulations that have the greatest potential to improve safety, improve efficiency, or reduce unnecessary burden. Both Options 2 and 3 can reasonably be expected to result in recommendations to make changes to Appendices A and B.

The staff's approach to Option 2 is to reclassify SSCs that will be subject to special treatment in accordance with a "risk-informed safety class" (RISC) scheme [9, 10]. Provisions in Part 50, including the GDC and Appendix B, that require special treatment will be identified and modified as appropriate. The staff's approach essentially has two elements. The first is to replace the phrase "important to safety" with the phrase "risk-significant" (RISC-1 or RISC-2). The second element is to identify the special treatment(s) currently required for components "important to safety" and establish comparable special treatment requirements for components that are "risk- significant."

The staff's work on Option 3, risk-informing the technical requirements in Part 50, is founded on a framework that establishes the approach, process, and guidelines to be applied in reviewing, formulating, and recommending risk-informed alternatives [11, 12, 13]. That framework is directed toward existing regulations that have an impact on preventing or mitigating core damage accidents. SECY-00-0198 identifies five key features of the framework, as follows:

(1) Utilize a risk-informed, defense-in-depth approach, building on the principles in Regulatory Guide 1.174, the reactor oversight cornerstones, and recommendations to the Commission in the ACRS report dated May 19, 1999.

(2) Include in the defense-in-depth approach elements that are dependent on risk insights, as well as elements that are independent of risk insights. Risk insights are used to set guidelines that limit the frequency of initiating events, core damage, radionuclide release, and public health effects. Defense-in-depth elements employed independent of risk insights include maintaining balance between prevention and mitigation, and maintaining the defense-in-depth objectives of the current GDC.

(3) Consider both design-basis and core-melt accidents.

(4) Consider uncertainties.

(5) Use the safety goals, together with the subsidiary objectives of 10^{-4} per reactor year for CDF and 10^{-5} per reactor year for LERF, to define how safe is safe enough.

The Option 3 framework (pp. 2–3, Attachment 1, SECY-00-0198) lists one of the elements of the defense-in-depth strategy as maintaining the defense-in-depth objectives of the current GDC. Coarse screening of Part 50 identifies the GDC and QAC as potentially relevant to risk-informing. The staff considers that the scope of Option 3 should focus on ensuring balance among risk, defense-in-depth, and safety margins, and applying safety principles established in RG 1.174. Candidates for change include specific requirements in Part 50, design-basis accidents, and regulatory guides. Specific requirements may be added, modified, or deleted.

5. OPTIONS FOR RISK-INFORMING THE GENERAL DESIGN CRITERIA

There are essentially three ways to approach risk-informing the GDC. The first, and probably the simplest, is to modify the scope of the GDC so that they apply only to SSCs that are risk-significant. The second approach is to examine the individual criteria and modify the requirements themselves to reflect risk as the appropriate measure of safety. As currently written, the GDC are structured around the design-basis accident model for safety regulation. The emphasis is on compliance with specific requirements that are intended to ensure no undue risk. The third approach is to replace the GDC with high-level design objectives and risk acceptance criteria. These risk acceptance criteria could be the Commission's current safety goals plus subsidiary objectives, such as CDF and LERF. Numerical values need not be incorporated in the regulations themselves, but the risk metrics to be used should be identified. The current GDC, or appropriate modifications thereof, could then be moved from the regulations themselves into guidance documents.

5.1 Modifying the Scope

Modifying the scope of the GDC to address risk-significant SSCs rather than those that are important to safety is the simplest approach. This is essentially the path that the NRC staff has chosen in its work to risk-inform the special treatment requirements in Part 50. It is also similar to an informal proposal that the NEI made for discussion purposes a few years ago. The scope of 13 of the GDC is "structures, systems, and components important to safety." NEI's suggestion was to define the phrase "important to safety" to be consistent with the idea of "a structure, system, or component which operating experience or probabilistic risk assessment has shown to be significant to public health and safety."

A more direct modification is suggested by the NRC staff's approach to Option 2. Specifically, where the phrase "important to safety" appears in the criteria, it could be replaced by "important to risk." "Important to risk" could then be defined as those SSCs included in RISC-1 or RISC-2 [9]. The criteria that could be so modified are indicated by a "1" in Column 8 of Table 1 to this report.

The risk focus of six of the GDC might also be improved by redefining the design-basis loss-of-coolant accident (LOCA). Criterion 38, for example, requires provision of containment heat removal for "any loss-of-coolant accident," and further requires the system to accomplish its safety function "assuming a single failure." Appendix A defines LOCAs to include "the double-ended rupture of the largest pipe of the reactor coolant system." NEI has suggested defining LOCAs in terms of a "likely pipe break size spectrum based on risk insights and operating experience." The NRC staff has indicated a willingness to discuss redefining the design basis LOCA, with the expectation that full system capacity, redundancy, and reliability would be required to accommodate the most likely break size, but not for a double-ended break of the largest pipe. Criterion 4, Environmental and Dynamic Effects Design Bases, already provides that "dynamic effects associated with postulated pipe ruptures in nuclear power units may be excluded from the design basis when analyses reviewed and approved by the Commission demonstrate that the probability of fluid system piping rupture is extremely low under conditions consistent with the design basis for the piping." Criteria that might be modified by changing the definition of the design-basis LOCA are indicated by a "2" in Column 8 of Table 1 to this report.

5.2 Modifying Individual Requirements

Aside from the scope and definition changes discussed above, the diversity of requirements established by the GDC makes it difficult, with perhaps one exception, to identify broad classes of proposed modifications. One possible change that affects several criteria is to replace the "single failure criterion" with a quantitative safety function reliability requirement. A single failure is defined in Appendix A as "... an occurrence which results in the loss of capability of a component to perform its intended safety functions." Criterion 17, for example, then requires that "... onsite electrical power supplies ... shall have sufficient independence, redundancy, and testability to perform their safety functions assuming a single failure." Similar provisions are included in seven other criteria, as indicated by the notation "SFC" in Column 9 of Table 1 to this report. The objective of the SFC is obviously to achieve a very high reliability for the safety-significant SSCs. Current probabilistic risk analysis methods should permit a reliability goal to be established for any particular safety function, consistent with its importance in reducing risk. The system designer can then provide diversity, redundancy, testability, or other design features necessary to meet the reliability goal. The final design configuration need not meet the SFC if the required reliability can be achieved without meeting it.

There are at least two approaches to revising individual technical requirements of the GDC. One approach involves what might be termed "incremental change." In this case, the individual criteria might be modified as necessary to directly address risk-significant issues. As written, the GDC are intended to reduce risk, but both the sources of risk and the means of reducing risk are assumed. Criterion 41, for example, requires provision of a containment atmosphere cleanup system, and further requires that redundant features ensure that the safety function can be accomplished assuming that either offsite or onsite power is unavailable and also assuming a single failure. The underlying assumptions are that reducing fission product and combustible gas concentrations will reduce risk, and that the required system reliability can be accomplished by providing the specified redundancy. Current probabilistic risk assessment (PRA) methods should allow system reliability requirements to be established on the basis of the system's contribution to achieving an acceptable LERF. The system designer could then provide the design features

11

required to achieve the necessary reliability, which might or might not be the features specified in Criterion 41 as it is currently written. The net effect of this incremental approach would be to preserve both the structure of the GDC and the safety functions and features that they address. The risk focus of some criteria would be improved, while other criteria (in particular, the defense-in-depth requirements) would be left unchanged. This approach is similar to that proposed by the NRC staff in its work to risk-inform the technical requirements of 10 CFR Part 50 (Option 3).

A second approach to risk-informing the GDC is to rewrite them completely in terms of risk, essentially starting with a blank sheet of paper. Design features and phenomenology that are understood to be important in achieving acceptably low CDF or LERF could be addressed in general terms. The designer would be free to provide whatever specific design features were needed to meet the required risk metrics. Criteria would not be included for design features or phenomena that PRA has shown to have little impact on reducing risk. Defense-in-depth issues would be addressed directly by requiring, for example, an emergency core cooling system (ECCS) and a containment. This approach is obviously a major undertaking, and is not pursued in this paper. An important difference between this approach and the incremental approach described above is that the incremental approach would leave in place those criteria that "do no harm" from a risk standpoint. A complete risk-informed rewrite of the GDC would presumably eliminate all criteria that had no significant impact on risk.

5.3 Replacing General Design Criteria with Safety Goals

One step beyond rewriting the GDC in terms of risk would be to replace the GDC within the regulatory framework with high-level regulatory objectives (such as the Commission's safety goals) and risk acceptance criteria. One such framework has been suggested by ACRS member Thomas S. Kress. His thoughts are described in Attachment 3 to this report.

It should be noted that a framework such as that discussed in Attachment 3 does not eliminate the need for top-level (general) design criteria. Regulatory objectives and risk acceptance criteria must be translated into design criteria that tell the designer what systems and safety functions must be provided in order to meet the risk acceptance criteria. As such, GDC are a necessary element in the hierarchy of documents that govern the design. If they are not provided in the regulatory requirements, the design organization (or the regulator) must develop them from the regulatory objectives and risk acceptance criteria.

6. DISCUSSION OF SELECTED CRITERIA

Possible changes to individual criteria are best discussed by considering changes to selected criteria, and then identifying other criteria for which a similar change might be appropriate. The discussion in this section examines selected criteria, identifies possible changes to improve the focus on risk, and then identifies other criteria with similar provisions or structure. It should also be observed that the risk-significance of a particular criterion may not be obvious from reading it. In some cases, it may necessary to have knowledge of PRA results or, ideally, to have a PRA available to determine changes in risk metrics that might result if features required by a design criterion are added or deleted. The process of determining risk-significance by inspecting the words in the GDC carries with it some possibility of error.

Subsection headings in the discussion that follows correspond to the group titles used to classify the criteria in Appendix A to 10 CFR Part 50.

6.1 Group I, Overall Requirements

The first five criteria are designated "Overall Requirements." Criterion 1 reads as follows:

> Quality standards and records. Structures, systems, and components important to safety shall be designed, fabricated, erected, and tested to quality standards commensurate with the importance of the safety functions to be performed. Where generally recognized codes and standards are used, they shall be identified and evaluated to determine their applicability, adequacy, and sufficiency and shall be supplemented or modified as necessary to assure a quality product in keeping with the required safety function. A quality assurance program shall be established and implemented in order to provide adequate assurance that these structures, systems, and components will satisfactorily perform their safety functions. Appropriate records of the design, fabrication, erection, and testing of structures, systems, and components important to safety shall be maintained by or under the control of the nuclear power unit licensee throughout the life of the unit.

This criterion does not appear to contain anything that is inimical to risk-informed regulation, with the exception of the phrase "important to safety" and its definition in terms of "without undue risk" in the introduction to Appendix A. As noted earlier, "important to safety" could be replaced by "important to risk," which could then be defined to mean SSCs in the RISC-1 and RISC-2 classes described in SECY-99-256 and SECY-00-0194. Obviously, other definitions of "important to risk" could be used and incorporated into the introduction to Appendix A. No other changes would need to be made for this particular criterion, and the same change can be made to the scope of Criteria 2, 3, and 4. Individual criteria where changes of this type are appropriate are indicated by the word "scope" in Column 9 of Table 1 to this report.

Criterion 5, which is also included in the "Overall Requirements" group, presents a somewhat different situation. This criterion reads as follows:

> Sharing of structures, systems, and components. Structures, systems, and components important to safety shall not be shared among nuclear power units unless it can be shown that such sharing will not significantly impair their ability to perform their safety functions, including, in the event of an accident in one unit, an orderly shutdown and cooldown of the remaining units.

13

As with the first four criteria, the scope of Criterion 5 can be changed from "important to safety" to "important to risk." Beyond that, it can be noted that a decision to share or not share systems is the kind of issue that quantitative risk assessment can address directly. PRA should be able to answer the question of what systems should be shared or interconnected between units to reduce CDF. At a minimum, risk-informing this criterion should change the bias so as to encourage the designer to identify the configurations that will reduce risk. As currently worded, the criterion can be satisfied by not sharing systems among units, which may not be the best decision from the standpoint of reducing risk.

6.2 Group II. Protection by Multiple Fission Product Barriers

This group encompasses 10 criteria, and the group title does not seem to fit well because the criteria vary in their scope, content, and structure. Criterion 10 reads as follows:

> Reactor design. The reactor core and associated coolant, control, and protection systems shall be designed with appropriate margin to assure that specified acceptable fuel design limits are not exceeded during any condition of normal operation, including the effects of anticipated operational occurrences.

As worded, Criterion 10 does not appear to inhibit risk-informed design, and probably need not be changed. The same conclusion is appropriate for Criterion 11, Reactor Inherent Protection.

Similarly, Criterion 12, Suppression of Reactor Power Oscillations, does not appear to inhibit risk-informed regulation. It probably makes little contribution to either improving safety or reducing risk, however. It reads as follows:

> Suppression of reactor power oscillations. The reactor core and associated coolant, control, and protection systems shall be designed to assure that power oscillations which can result in conditions exceeding specified acceptable fuel design limits are not possible or can be reliably and readily detected and suppressed.

Criterion 10, by itself, would seem to prohibit power oscillations that might cause fuel design limits to be exceeded, making Criterion 12 superfluous.

Criterion 13, Instrumentation and Control, could be modified to improve the focus on risk. Criterion 13 reads as follows:

> Instrumentation and control. Instrumentation shall be provided to monitor variables and systems over their anticipated ranges for normal operation, for anticipated operational occurrences, and for accident conditions as appropriate to assure adequate safety, including those variables and systems that can affect the fission process, the integrity of the reactor core, the reactor coolant pressure

14

boundary, and the containment and its associated systems. Appropriate controls shall be provided to maintain these variables and systems within prescribed operating ranges.

The term "adequate safety" could be replaced by "acceptably low risk" to avoid the implied "adequate protection" standard. Risk considerations could also be emphasized by specific mention of instrumentation needed to implement emergency operating procedures and severe accident management guidelines. Note that although this criterion was not changed following the accident at Three Mile Island, its implementation was changed substantially. Specifically, monitoring of additional variables was required by the addition of post-accident sampling systems, safety parameter display systems, and pressurized-water reactor (PWR) vessel level indication systems.

Criteria 14 and 15 also do not appear to be inimical to risk-informing Part 50. Criterion 14, for example, reads as follows:

> Reactor coolant pressure boundary. The reactor coolant pressure boundary shall be designed, fabricated, erected, and tested so as to have an extremely low probability of abnormal leakage, of rapidly propagating failure, and of gross rupture.

Criterion 15, Reactor Coolant System Design, requires the following:

> Reactor coolant system design. The reactor coolant system and associated auxiliary, control, and protection systems shall be designed with sufficient margin to assure that the design conditions of the reactor coolant pressure boundary are not exceeded during any condition of normal operation, including anticipated operational occurrences.

Neither of these criteria would appear to be a barrier to either risk-informed design or risk-informed inspection and testing.

Criterion 16, Containment Design, has a number of risk implications. This criterion reads as follows:

> Containment design. Reactor containment and associated systems shall be provided to establish an essentially leak-tight barrier against the uncontrolled release of radioactivity to the environment and to assure that the containment design conditions important to safety are not exceeded for as long as postulated accident conditions require.

As written, Criterion 16 requires that containment be provided, that it be essentially leak tight, and that conditions "important to safety" not be exceeded. Again, the most obvious change to address risk is to replace "important to safety" with "important to risk." The requirement for a containment is clearly a defense-in-depth measure. It is not obvious, however, that requiring an "essentially leak-tight barrier" directs the designer to a containment design that is most effective in reducing risk, since early containment failure dominates severe accident risk. Leakage is less of a concern from a risk standpoint than is containment bypass or gross structural failure. To be risk-informed, this criterion should probably deal directly with conditions that might result from core melt accidents, and focus on reducing the probability of early containment failure. (The solution is not clear at this time, but the problem is worth noting.)

Criterion 17, Electrical Power, also has a number of risk implications. The criterion reads as follows:

> Electrical power. An onsite electric power system and an offsite electric power system shall be provided to permit functioning of structures, systems, and components important to safety. The safety function for each system (assuming the other system is not functioning) shall be to provide sufficient capacity and capability to assure that (1) specified acceptable fuel design limits and design conditions of the reactor coolant pressure boundary are not exceeded as a result of anticipated operational occurrences and (2) the core is cooled and containment integrity and other vital functions are maintained in the event of postulated accidents.

> The onsite electric power supplies, including the batteries, and the onsite electric distribution system, shall have sufficient independence, redundancy, and testability to perform their safety functions assuming a single failure.

> Electric power from the transmission network to the onsite electric distribution system shall be supplied by two physically independent circuits (not necessarily on separate rights of way) designed and located so as to minimize to the extent practical the likelihood of their simultaneous failure under operating and postulated accident and environmental conditions. A switchyard common to both circuits is acceptable. Each of these circuits shall be designed to be available in sufficient time following a loss of all onsite alternating current power supplies and the other offsite electric power circuit, to assure that specified acceptable fuel design limits and design conditions of the reactor coolant pressure boundary are not exceeded. One of these circuits shall be designed to be available within a few seconds following a loss-of-coolant accident to assure that core cooling, containment integrity, and other vital safety functions are maintained.

> Provisions shall be included to minimize the probability of losing electric power from any of the remaining supplies as a result of, or coincident with, the loss of power generated by the nuclear power unit, the loss of power from the

16

transmission network, or the loss of power from the onsite electric power supplies.

The objective of Criterion 17 is to provide reliable electric power for vital components. Again, the scope can be made risk-informed by changing "important to safety" to "important to risk." Beyond that, the aggregate of all individual requirements is to ensure reliability. There are a number of assumptions embedded in this criterion regarding sources of risk and effective means of reducing risk. The criterion can only be met by providing both onsite and offsite electrical power systems. The onsite system(s) must be capable of performing the intended safety functions assuming a single failure. The offsite system(s) must consist of two physically independent circuits, one of which "must be available within a few seconds following a loss-of-coolant accident." A common switchyard is deemed acceptable, and separate rights of way are not required. The redundancy and independence required are intended to provide a success path for safety functions even if multiple failures occur.

Current probabilistic analysis techniques should provide the capability to demonstrate compliance with a specified overall reliability goal for electric power, without the need to require diversity and redundancy at the level of specificity used in Criterion 17. In particular, the requirement that safety functions be performed "assuming a single failure" is not necessary if a specified, acceptable reliability objective is achieved, and results in acceptably low core damage frequency or large release frequency. Also, it can probably be shown that the requirement for a circuit to be available "within a few seconds following a loss-of-coolant accident" is not necessary from a risk standpoint.

Seven other criteria, as indicated in Column 9 of Table 1 of this report, require that safety functions be accomplished "assuming a single failure." Each could be restated in terms of a reliability goal required to achieve specified acceptable risk goals, such as core damage frequency or large release frequency.

Criterion 18, Inspection and Testing of Electric Power Systems, supports Criterion 17 by requiring that electrical power systems must be designed to permit functional testing. Since the purpose of such testing is to ensure the reliability of the systems, testing requirements could be made part of the overall reliability goal. Testing requirements could be contingent on their necessity to meet the reliability objective. As indicated in Column 4 of Table 1 of this report, 15 criteria require testing, inspection, or the capability to inspect or test systems or functions that are mandated by other criteria. In virtually all cases, these requirements could be replaced by, or made part of, an overall system or function reliability goal. It should be noted, however, that for risk-important systems, the required high reliability probably cannot be achieved without periodic inspection and testing. Inspection and testing programs can be designed to be risk-informed. Criterion 18, as written, does not appear to preclude risk-informed testing and inspection and, thus, probably does not require revision to support risk-informed regulation. The same observation is true for other criteria that require inspection or testing.

17

The last criterion in Group II is Criterion 19, Control Room, which states, in part:

> Control room. A control room shall be provided from which actions can be taken to operate the nuclear power unit safely under normal conditions and to maintain it in a safe condition under accident conditions, including loss-of-coolant accidents. Adequate radiation protection shall be provided to permit access and occupancy of the control room under accident conditions without personnel receiving radiation exposures in excess of 5 rem whole body, or its equivalent to any part of the body, for the duration of the accident.
>
> Equipment at appropriate locations outside the control room shall be provided (1) with a design capability for prompt hot shutdown of the reactor, including necessary instrumentation and controls to maintain the unit in a safe condition during hot shutdown, and (2) with a potential capability for subsequent cold shutdown of the reactor through the use of suitable procedures.

The risk focus of this criterion could be improved if the LOCA were defined in terms of a likely break size, rather than the double-ended break of the largest pipe. The second paragraph, which requires certain equipment outside the control room, is essentially a defense-in-depth requirement. It seems likely that probabilistic risk analysis could be used to determine whether this requirement has a significant risk benefit.

6.3 Group III, Protection and Reactivity Control Systems

Of the 10 criteria in this group, 6 (Criteria 20, 22, 23, 27, 28 and 29) do not appear to be inimical to risk-informed regulation, and probably need not be changed to risk-inform the GDC. Of the remaining four, three (Criteria 21, 24, and 25) require that safety functions must be accomplished "assuming a single failure," and could be revised by establishing a reliability goal for the success of the safety function, rather than specifically requiring redundancy.

Criterion 20 reads as follows:

> Protection system functions. The protection system shall be designed (1) to initiate automatically the operation of appropriate systems including the reactivity control systems, to assure that specified acceptable fuel design limits are not exceeded as a result of anticipated operational occurrences and (2) to sense accident conditions and to initiate the operation of systems and components important to safety.

As written, Criterion 20 probably requires no change to be compatible with risk-informing Part 50. Its primary purpose is to require automatic initiation of safety functions as opposed to relying on operator action. The risk focus could be improved by changing "important to safety" to "important to risk" where it requires the protection system to "initiate the operation of systems and components important to safety."

Criterion 21, Protection System Reliability and Testability, reads as follows:

> Protection system reliability and testability. The protection system shall be designed for high functional reliability and inservice testability commensurate with the safety functions to be performed. Redundancy and independence designed into the protection system shall be sufficient to assure that (1) no single failure results in loss of the protection function and (2) removal from service of any component or channel does not result in loss of the required minimum redundancy unless the acceptable reliability of operation of the protection system can be otherwise demonstrated. The protection system shall be designed to permit periodic testing of its functioning when the reactor is in operation, including a capability to test channels independently to determine failures and losses of redundancy that may have occurred.

The obvious purpose of this criterion is high reliability of the protection system. The underlying assumption is that the desired reliability can best be achieved by redundancy and independence. The wording also suggests the possibility that "acceptable reliability of operation of the protection system can be otherwise demonstrated." Current analysis capabilities should permit establishing a quantitative protection system reliability goal, and then demonstrating that it has been achieved and maintained without explicitly invoking the "single failure" requirement.

Criterion 22 addresses protection system independence, as follows:

> Protection system independence. The protection system shall be designed to assure that the effects of natural phenomena, and of normal operating, maintenance, testing, and postulated accident conditions on redundant channels do not result in loss of the protection function, or shall be demonstrated to be acceptable on some other defined basis. Design techniques, such as functional diversity or diversity in component design and principles of operation, shall be used to the extent practical to prevent loss of the protection function.

As written, Criterion 22 has some implied flexibility ("... or shall be demonstrated to be acceptable on some other defined basis"), and probably need not be revised from the standpoint of risk-informing Part 50. The same is true for Criterion 23, Protection System Failure Modes.

Criteria 24 and 25 don't appear to do any particular harm from a risk perspective, but they could also be made part of an overall reliability goal for the protection function. They currently read as follows:

Criterion 24—Separation of protection and control systems. The protection system shall be separated from control systems to the extent that failure of any single control system component or channel, or failure or removal from service of any single protection system component or channel which is common to the control and protection systems leaves intact a system satisfying all reliability, redundancy, and independence requirements of the protection system. Interconnection of the protection and control systems shall be limited so as to assure that safety is not significantly impaired.

Criterion 25—Protection system requirements for reactivity control malfunctions. The protection system shall be designed to assure that specified acceptable fuel design limits are not exceeded for any single malfunction of the reactivity control systems, such as accidental withdrawal (not ejection or dropout) of control rods.

The specificity in these criteria with respect to redundancy and independence is probably not necessary given current reliability analysis capability. Similarly, the specification of limited interconnection may not be necessary in light of current reliability analysis methods. Criterion 25 contains the implicit judgement that control rod ejection or dropout is a sufficiently improbable event that some fuel damage may be acceptable. Other, presumably more probable, malfunctions of the reactivity control system, should not result in exceeding specified acceptable fuel design limits, which are understood to preclude fuel damage. It seems likely that these criteria could be rewritten in terms of protection and control system reliability goals, without directly invoking the single failure requirement.

Criterion 26, Reactivity Control System Redundancy and Capability, could probably be rewritten to take advantage of current reliability analysis methods. It currently reads as follows:

Reactivity control system redundancy and capability. Two independent reactivity control systems of different design principles shall be provided. One of the systems shall use control rods, preferably including a positive means for inserting the rods, and shall be capable of reliably controlling reactivity changes to assure that under conditions of normal operation, including anticipated operational occurrences, and with appropriate margin for malfunctions such as stuck rods, specified acceptable fuel design limits are not exceeded. The second reactivity control system shall be capable of reliably controlling the rate of reactivity changes resulting from planned, normal power changes (including xenon burnout) to assure acceptable fuel design limits are not exceeded. One of the systems shall be capable of holding the reactor core subcritical under cold conditions.

As written, Criterion 26 is a bit convoluted, and it is not immediately clear why it is structured as it is. Its structure is probably an endorsement of the reactivity control systems adopted by PWR and BWR designers by the mid-1960s. For PWRs, the two independent reactivity control systems are control rods and soluble boron addition to the moderator. For BWRs, the two systems are recirculation flow control and control rods. Criterion 26 also appears to assign specific functions to each of two systems. "One system shall use control rods and ... shall ... assure that under conditions of normal operation ... fuel design limits are not exceeded. The second reactivity control system shall be capable of ... controlling ... reactivity changes ... from planned, normal power changes." Why the language appears to differentiate between "normal operation" and "normal power changes" is not clear. It might also be observed that the anticipated transient without scram (ATWS) rule (10 CFR 50.62) later established substantial additional redundancy and capability requirements for reactivity control. It is not immediately obvious why the ATWS rule did not include a change to Criterion 26.

The remaining criteria in this group, including Criterion 27 (Combined Reactivity Control Systems Capability), Criterion 28 (Reactivity Limits), and Criterion 29 (Protection Against Anticipated Operational Occurrences), probably do not require any revision to support risk-informed regulation.

6.4 Group IV, Fluid Systems

This group encompasses 17 criteria, of which 3 address the integrity of the reactor coolant pressure boundary, and 14 address functions related to coolant inventory control, residual heat removal, emergency core cooling, containment atmosphere cleanup, containment heat removal, and ultimate heat sink. The latter subset includes three criteria, namely Criterion 35 (Emergency Core Cooling), Criterion 38 (Containment Heat Removal), and Criterion 41 (Containment Atmosphere Cleanup), that are defense-in-depth provisions.

Criterion 30 reads as follows:

> Quality of reactor coolant pressure boundary. Components which are part of the reactor coolant pressure boundary shall be designed, fabricated, erected, and tested to the highest quality standards practical. Means shall be provided for detecting and, to the extent practical, identifying the location of the source of reactor coolant leakage.

As written, Criterion 30 does not appear to impede risk-informing Part 50. Similarly, the two following supporting criteria (Fracture Prevention of the Reactor Coolant Pressure Boundary, and Inspection of the Reactor Coolant Pressure Boundary) should not require any changes.

Criterion 33, Reactor Coolant Makeup, could possibly be improved from a risk-informed perspective. It currently reads as follows:

> Reactor coolant makeup. A system to supply reactor coolant makeup for protection against small breaks in the reactor coolant pressure boundary shall be provided. The system safety function shall be to assure that specified acceptable fuel design limits are not exceeded as a result of reactor coolant loss due to

leakage from the reactor coolant pressure boundary and rupture of small piping or other small components which are part of the boundary. The system shall be designed to assure that for onsite electric power system operation (assuming offsite power is not available) and for offsite electric power system operation (assuming onsite power is not available), the system safety function can be accomplished using the piping, pumps, and valves used to maintain coolant inventory during normal reactor operation.

The purpose of Criterion 33 is to provide a high-reliability, high-availability system to respond to small breaks, which are (apparently) assumed to be more frequent than the larger breaks to which the ECCS is intended to respond. The underlying assumption is that providing this particular system will reduce core damage frequency. Current PRA methods could be used to determine whether the requirements in this criterion are more effective in reducing CDF than some alternative requirements. For example, is the requirement to use "... piping, pumps, and valves used to maintain coolant inventory during normal reactor operation" an important contributor to reducing risk? The criterion also requires the safety function to succeed if either onsite or offsite power is unavailable. That provision would appear to be risk-significant, but PRA could be used to confirm that significance.

Criterion 34, Residual Heat Removal, also could be modified to improve its impact on risk. It currently reads as follows:

Residual heat removal. A system to remove residual heat shall be provided. The system safety function shall be to transfer fission product decay heat and other residual heat from the reactor core at a rate such that specified acceptable fuel design limits and the design conditions of the reactor coolant pressure boundary are not exceeded.

Suitable redundancy in components and features, and suitable interconnections, leak detection, and isolation capabilities shall be provided to assure that for onsite electric power system operation (assuming offsite power is not available) and for offsite electric power system operation (assuming onsite power is not available) the system safety function can be accomplished, assuming a single failure.

Intuitively, one expects the function addressed here to be very risk-significant. The most straightforward change is to replace the "single failure" requirement with an overall system reliability goal, such that an acceptably low CDF can be achieved. There is also an imbedded defense-in-depth provision, which combines the assumed single failure with the assumed unavailability of either offsite or onsite power. This provision might also be subsumed into the overall reliability goal. (Similar requirements appear in Criteria 35, 38, 41, and 44.)

Criterion 35, Emergency Core Cooling, reads as follows:

> Emergency core cooling. A system to provide abundant emergency core cooling shall be provided. The system safety function shall be to transfer heat from the reactor core following any loss of reactor coolant at a rate such that (1) fuel and clad damage that could interfere with continued effective core cooling is prevented and (2) clad metal-water reaction is limited to negligible amounts.

> Suitable redundancy in components and features, and suitable interconnections, leak detection, isolation, and containment capabilities shall be provided to assure that for onsite electric power system operation (assuming offsite power is not available) and for offsite electric power system operation (assuming onsite power is not available) the system safety function can be accomplished, assuming a single failure.

The requirement to provide an ECCS, independent of extensive measures to prevent LOCAs from occurring, is a defense-in-depth measure. As written and implemented, Criterion 35 treats the double-ended break of the largest pipe in the reactor coolant system as the design-basis for the ECCS. Current PRA methods allow the designer to identify the most probable break sizes that may be larger contributors to core damage than the design-basis LOCA. The NRC staff and the nuclear industry are currently discussing revision of this criterion to better address the breaks that contribute most to risk, and to relax requirements for less likely breaks. The current discussions contemplate, for example, not requiring two full-capacity ECCS trains to respond to the double-ended break of the largest pipe.

Criterion 35 is supported by Criteria 36 and 37, Inspection of Emergency Core Cooling System and Testing of Emergency Core Cooling System. The criteria on inspection and testing, of course, are intended to enhance the reliability of the ECCS. These three criteria could be considered together to establish capacity and reliability requirements to achieve an acceptably low CDF.

The next three criteria deal with the provision, testing, and inspection of a containment heat removal system. Criterion 38 reads as follows:

> Containment heat removal. A system to remove heat from the reactor containment shall be provided. The system safety function shall be to reduce rapidly, consistent with the functioning of other associated systems, the containment pressure and temperature following any loss-of-coolant accident and maintain them at acceptably low levels.

> Suitable redundancy in components and features, and suitable interconnections, leak detection, isolation, and containment capabilities shall be provided to assure that for onsite electric power system operation (assuming offsite power is not

available) and for offsite electric power system operation (assuming onsite power is not available) the system safety function can be accomplished, assuming a single failure.

To the extent that containment heat removal reduces the probability of containment failure, one would expect it to contribute to reducing risk. Making use of current probabilistic analysis methods, it may be possible to improve the risk focus by combining this criterion with the criteria on inspection and testing, and replacing the "single failure" requirement with an overall system reliability goal. The designer could then provide whatever system features were necessary to meet the reliability goal. This also may be a case where, from a risk standpoint, redundancy requirements can be reduced for relatively unlikely large break sizes.

The next three criteria establish requirements for provision, testing, and inspection of a containment atmosphere cleanup system. Criterion 41 reads as follows:

> Containment atmosphere cleanup. Systems to control fission products, hydrogen, oxygen, and other substances which may be released into the reactor containment shall be provided as necessary to reduce, consistent with the functioning of other associated systems, the concentration and quality of fission products released to the environment following postulated accidents, and to control the concentration of hydrogen or oxygen and other substances in the containment atmosphere following postulated accidents to assure that containment integrity is maintained.
>
> Each system shall have suitable redundancy in components and features, and suitable interconnections, leak detection, isolation, and containment capabilities to assure that for onsite electric power system operation (assuming offsite power is not available) and for offsite electric power system operation (assuming onsite power is not available) its safety function can be accomplished, assuming a single failure.

The discussion of Criterion 38, Containment Heat Removal, above, applies equally well to Criterion 41. It seems likely, however, that containment atmosphere cleanup would be somewhat less risk-significant than containment heat removal, since the latter bears more directly on reducing pressure and temperature loads on containment.

The next three criteria establish requirements for provision, inspection, and testing of a cooling water system and ultimate heat sink. Criterion 44 reads as follows:

> Cooling water. A system to transfer heat from structures, systems, and components important to safety, to an ultimate heat sink shall be provided. The system safety function shall be to transfer the combined heat load of these SSCs under normal operating and accident conditions.

24

Suitable redundancy in components and features, and suitable interconnections, leak detection, and isolation capabilities shall be provided to assure that for onsite electric power system operation (assuming offsite power is not available) and for offsite electric power system operation (assuming onsite power is not available) the system safety function can be accomplished, assuming a single failure.

Again, the general comments regarding the criteria for containment heat removal are applicable here. One expects that the ability to transfer thermal loads to the ultimate heat sink is highly risk-significant.

6.5 Group V, Reactor Containment

This group comprises eight criteria addressing containment design basis, containment testing, and containment penetrations. The requirement to provide a containment is a defense-in-depth issue. As noted earlier, the containment designs resulting from application of the current GDC may not be the best designs from a risk standpoint. It seems reasonable to expect that PRA methods could be used to identify those criteria that have a significant impact on risk, and those that have a small or even counter-productive impact. Criterion 50 reads as follows:

> Containment design basis. The reactor containment structure, including access openings, penetrations, and the containment heat removal system shall be designed so that the containment structure and its internal compartments can accommodate, without exceeding the design leakage rate and with sufficient margin, the calculated pressure and temperature conditions resulting from any loss-of-coolant accident. This margin shall reflect consideration of (1) the effects of potential energy sources which have not been included in the determination of the peak conditions, such as energy in steam generators and as required by §50.44 energy from metal-water and other chemical reactions that may result from degradation but not total failure of emergency core cooling functioning, (2) the limited experience and experimental data available for defining accident phenomena and containment responses, and (3) the conservatism of the calculational model and input parameters.

Criterion 51, which concerns fraction prevention, reads as follows:

> Fracture prevention of containment pressure boundary. The reactor containment boundary shall be designed with sufficient margin to assure that under operating, maintenance, testing, and postulated accident conditions (1) its ferritic materials behave in a nonbrittle manner and (2) the probability of rapidly propagating fracture is minimized. The design shall reflect consideration of service temperatures and other conditions of the containment boundary material during operation, maintenance, testing, and postulated accident conditions, and the

uncertainties in determining (1) material properties, (2) residual, steady state, and transient stresses, and (3) size of flaws.

Criteria 50 and 51, together with the seven subsequent criteria, amplify the basic containment requirement established in Criterion 16. If the basic requirement for an essentially leak-tight containment is maintained, Criteria 50 and 51 probably need not be revised. Criterion 50 could be risk-informed by redefining the design-basis LOCA, so that full redundancy and diversity of containment heat removal systems is not required for relatively unlikely large pipe breaks.

Criteria 52 and 53 provide for leak testing and inspection, and ensure that the containment will fulfill its function of providing an essentially leak-tight barrier against fission product release. Testing and inspection programs can be designed to be risk-informed, and these criteria probably do not require any changes. If the current understanding is correct, (i.e., if risk is dominated by early containment failure or containment bypass scenarios), these two criteria may not be particularly risk-significant.

Criteria 54 through 57 establish isolation requirements for systems that penetrate containment. Criterion 54 reads as follows:

> Piping systems penetrating containment. Piping systems penetrating primary reactor containment shall be provided with leak detection, isolation, and containment capabilities having redundancy, reliability, and performance capabilities which reflect the importance to safety of isolating these piping systems. Such piping systems shall be designed with a capability to test periodically the operability of the isolation valves and associated apparatus and to determine if valve leakage is within acceptable limits.

The risk focus of this criterion could be improved by replacing "importance to safety" with "importance to risk." If severe accident risk is dominated by containment bypass or large early release, this criterion may not be important from a risk standpoint.

Criteria 55, 56, and 57 provide a great deal of specificity regarding the isolation of systems that penetrate the containment boundary. Criterion 55 states, for example:

> Reactor coolant pressure boundary penetrating containment. Each line that is part of the reactor coolant pressure boundary and that penetrates primary reactor containment shall be provided with containment isolation valves as follows, unless it can be demonstrated that the containment isolation provisions for a specific class of lines, such as instrument lines, are acceptable on some other defined basis:
>
> (1)   One locked closed isolation valve inside and one locked closed isolation valve outside containment; or

(2)   One automatic isolation valve inside and one locked closed isolation valve outside containment; or

(3)   One locked closed isolation valve inside and one automatic isolation valve outside containment. A simple check valve may not be used as the automatic isolation valve outside containment; or

(4)   One automatic isolation valve inside and one automatic isolation valve outside containment. A simple check valve may not be used as the automatic isolation valve outside containment.

Isolation valves outside containment shall be located as close to containment as practical and upon loss of actuating power, automatic isolation valves shall be designed to take the position that provides greater safety.

Other appropriate requirements to minimize the probability or consequences of an accidental rupture of these lines or of lines connected to them shall be provided as necessary to assure adequate safety. Determination of the appropriateness of these requirements, such as higher quality in design, fabrication, and testing, additional provisions for inservice inspection, protection against more severe natural phenomena, and additional isolation valves and containment, shall include consideration of the population density, use characteristics, and physical characteristics of the site environs.

The importance of these requirements from a risk standpoint can probably only be determined by using PRA methods to determine the risk impact of providing, or not providing, the specified isolation features. Containment isolation provisions, including the location of system isolation valves, may or may not reflect the best design features from a risk standpoint. One suspects that the focus on risk suffers when the degree of specificity in the design criteria increases.

6.6 Group VI, Fuel and Radioactivity Control

This group includes five criteria. As a group, they are probably not very risk-significant because they deal primarily with monitoring and controlling relatively small releases. The one obvious exception is the requirement to monitor radioactivity releases to the plant environs, which provides important information for emergency response to reduce risk. Specifically, Criterion 60 reads as follows:

Control of releases of radioactive materials to the environment. The nuclear power unit design shall include means to control suitably the release of radioactive materials in gaseous and liquid effluents and to handle radioactive solid wastes produced during normal reactor operation, including anticipated operational occurrences. Sufficient holdup capacity shall be provided for retention of gaseous and liquid effluents containing radioactive materials, particularly where unfavorable site environmental conditions can be expected to impose unusual operational limitations upon the release of such effluents to the environment.

This criterion requires controls on the routine release of radioactive material resulting from normal operations. As such, it is intended to reduce chronic exposure from plant operations, rather than to address severe accident risk. Given that intent, it is probably a good example of a criterion that is important from the standpoint of keeping radiation exposures as low as practicable, and its usefulness could be diminished by making it "risk-informed."

Criterion 61, Fuel Storage and Handling and Radioactivity Control, is a mixture of provisions for worker safety and potentially important public risk considerations. It currently reads as follows:

Fuel storage and handling and radioactivity control. The fuel storage and handling, radioactive waste, and other systems which may contain radioactivity shall be designed to assure adequate safety under normal and postulated accident conditions. These systems shall be designed (1) with a capability to permit appropriate periodic inspection and testing of components important to safety, (2) with suitable shielding for radiation protection, (3) with appropriate containment, confinement, and filtering systems, (4) with a residual heat removal capability having reliability and testability that reflects the importance to safety of decay heat and other residual heat removal, and (5) to prevent significant reduction in fuel storage coolant inventory under accident conditions.

The risk focus of Criterion 61 could be improved by replacing "adequate safety" with "acceptably low risk," and "importance to safety" with "importance to risk." No other change appears necessary.

Criterion 62, which concerns criticality prevention, is more an issue of a worker safety than public risk, but it is probably extremely important to the public perception of risk. It reads as follows:

Prevention of criticality in fuel storage and handling. Criticality in the fuel storage and handling system shall be prevented by physical systems or processes, preferably by use of geometrically safe configurations.

As written, Criterion 62 is not an impediment to risk-informing Part 50, and changes to make it risk-informed could diminish its value.

Criterion 63, Monitoring Fuel and Waste Storage, seems to overlap considerably with Criterion 61. Criterion 63 reads as follows:

Monitoring fuel and waste storage. Appropriate systems shall be provided in fuel storage and radioactive waste systems and associated handling areas (1) to detect conditions that may result in loss of residual heat removal capability and excessive radiation levels and (2) to initiate appropriate safety actions.

As written, Criterion 63 does not appear to be an impediment to risk-informed regulation.

Criterion 64 is a mixture of requirements addressing both occupational radiation safety concerns and public risk. It reads as follows:

Monitoring radioactivity releases. Means shall be provided for monitoring the reactor containment atmosphere, spaces containing components for recirculation of loss-of-coolant accident fluids, effluent discharge paths, and the plant environs for radioactivity that may be released from normal operations, including anticipated operational occurrences, and from postulated accidents.

As currently written, Criterion 64 is not inconsistent with risk-informed regulation.

7. CONCLUSIONS REGARDING RISK-INFORMING APPENDIX A

The general design criteria in Appendix A to 10 CFR Part 50 evolved during the 1960s, and represent the state-of-the-art at that time in designing light-water moderated and cooled reactors for safety. As such, they are structured to support the design-basis accident model of safety regulation. The underlying regulatory standard is "no undue risk to public health and safety," which is equivalent to the "adequate protection" standard derived from the Atomic Energy Act. The individual criteria vary greatly in their contribution to safety or reduction of risk. Collectively, they may be sufficient, in conjunction with the rest of the regulatory structure, to result in designs with acceptably low CDF or LERF. The validity of this speculation would be difficult to demonstrate, and it would even be difficult to determine the degree to which the current fleet of operating reactors meets the GDC.

As written, the GDC accomplish a number of design objectives. One such objective is stable, forgiving reactor designs with sufficient margin to avoid challenging safety limits during normal operations, including anticipated operational occurrences. Another objective is the provision of specific safety functions to provide defense-in-depth (i.e., limit accident initiators, terminate accident sequences quickly, and mitigate accidents that are not successfully terminated). Defense-in-depth is also specified through requirements for multiple fission product barriers, and the means to protect those barriers. The reliability of functions important to safety is ensured by specific requirements for independence, diversity, and redundancy in the system designs. One purpose of the GDC is to avoid getting into "uncharted territory," conditions for which it would be difficult to predict consequences or quantify the ability to avoid fuel damage (e.g., complex geometries beyond the designer's ability to calculate heat removal with confidence).

There are a number of ways to revise the GDC to be risk-informed or to remove impediments to risk-informing the rest of 10 CFR Part 50. The most straightforward option is to change the scope of certain criteria from "structures, systems, and components important to safety" to "structures, systems, and components important to risk." Such a change would redirect attention from those SSCs that are required to comply with design-basis accident requirements, to SSCs that are significant in achieving low risk metrics, such as CDF and LERF. Some components that are now governed by the GDC would no longer be, and other components that are currently classified as non-safety-related would then be covered by the GDC as "risk-significant." This is similar to the NRC staff's approach to risk-informing the special treatment requirements in Part 50 (Option 2).

Under Option 2, the staff is modifying the applicability of special treatment requirements on the basis of RISC classification of SSCs. In a similar manner, the GDC might be risk-informed by limiting their applicability to risk-important SSCs. Special treatment requirements are those requirements that are intended to provide a high degree of assurance that SSCs needed to withstand design-basis events will function under design-basis conditions. These requirements include equipment qualification, change control, documentation, reporting, maintenance, testing, surveillance, and quality assurance.

A second broad change would be to redefine the design-basis LOCA in terms of break sizes that are more likely than the double-ended rupture of the largest pipe in the system. PRA methods are capable of determining the most risk-significant break sizes and safety functions could be designed to reduce risk rather than to cope with the largest possible, but less likely, pipe break.

PRA methods can also be used to establish the quantitative reliability requirements for control, protection, and engineered safety features. Such requirements could replace the current, highly prescriptive requirements for independence, redundancy, and diversity, including revising or eliminating the SFC.

Finally, PRA methods could be used to determine the risk-significance of individual design criteria. If a given system or function were found to have little impact on risk, the related criterion could be eliminated or revised to improve its focus on risk.

While risk-informing the GDC is expected to be an important step in risk-informing Part 50, its impact will probably be limited unless supporting changes are made in the body of Part 50 and in relevant regulatory guidance documents. It seems evident that how one goes about risk-informing the GDC will depend on how the body of 10 CFR Part 50 is changed to be risk-informed.

If relatively radical changes in the regulatory framework can be considered, it would be possible to rewrite the GDC entirely in risk terms. Such an effort might also have the goal of making the resulting criteria independent of the particular reactor technology. Dr. Kress's proposal to replace GDC in the regulatory structure with safety goals and risk acceptance criteria also should be considered. Such an approach would not eliminate the need for GDC, but it would relocate them to a different place in the hierarchy of design documents. The reality is that designers must start somewhere, and almost never with a blank sheet of paper. Features that have been successfully used in the past to meet goals of any kind (cost, reliability, safety, ease of manufacture, rapid construction time, etc.) are likely to be used as a starting point for new designs.

8. APPLICABILITY OF APPENDIX A TO REACTORS OTHER THAN LWRs

The introduction to Appendix A states that "The General Design Criteria are also considered to be generally applicable to other types of nuclear power units and are intended to provide guidance in establishing the principal design criteria for such other units." The preceding sentence, however, says "These General Design Criteria establish minimum requirements for the principal design criteria for water-cooled nuclear power plants similar in design and location to plants for which construction permits have been issued by the Commission." In considering the applicability of particular criteria to nuclear power plants other than those "... similar in design and location to plants for which construction permits have been issued ...," a certain degree of caution is appropriate. Even those criteria that do not seem to be focused on water-cooled reactor technology may direct attention to phenomena that are less important to the safety of non-water-cooled reactors. Such criteria may also fail to direct attention to phenomena that are more important to the safety of non-water-cooled reactors.

As with the issue of risk-informing the GDC, the applicability of the GDC to non-water cooled reactors can best be dealt with one criterion at a time. Column 10 of Table 1 of this report shows a summary judgement regarding the applicability of each criterion to non-water-cooled reactors. The following paragraphs provide comments concerning selected criteria or groups of criteria. Broadly, each criterion can be classified according to whether it is applicable to all reactor types as written, its intent is applicable to all reactor types, or it does not apply to non-water-cooled reactors.

The criteria in the "Overall Requirements" group (Criteria 1 through 5) are applicable to all reactor types. In Group II, Protection by Multiple Fission Product Barriers, all but three criteria appear to be applicable to all reactor types. The intent of Criterion 13, Instrumentation and Control, should be applicable to all types, but the reference to containment may not be appropriate in all cases. Similarly, the intent of Criterion19, Control Room, should be universally applicable, but the reference to loss of coolant accidents may be misleading. The implication is that LOCAs are limiting for control room access and habitability. The limiting accidents in other reactor types may not be LOCAs, however. Criterion 16, which requires provision of an "essentially leak-tight" containment, is considered unnecessary by advocates of some gas- cooled reactor designs. If interest in the concept persists, there will be considerable discussion of what kind of containment, if any, is appropriate for gas-cooled reactors.

With only a few exceptions, the criteria in Group III, Protection and Reactivity Control Systems, seem to be applicable to all reactor types. Criterion 25, Protection System Requirements for Reactivity Control Malfunctions, contains a reference to control rods that may turn out to be inappropriate for some reactors. Criterion 28, Reactivity Limits, mentions a number of reactivity addition mechanisms (steam line rupture, cold water addition) that may not apply to non-water moderated reactors. Criteria 26 and 27, specifying reactivity control system capability, appear to be highly specific to water-cooled and moderated reactors as they are currently worded. The intent is applicable to all reactor types, but considerable rewording is probably necessary.

The criteria in Group IV range from applicable to all types, to not applicable to non-water-cooled reactors. Criteria 30, 31, and 32 deal with the integrity of the reactor coolant pressure boundary. While attention to that boundary is always important, there are some reactor types (pool types) for which other interfaces may be important. Consequently, the GDC should direct attention to the safety-critical interfaces (e.g., liquid metal-to-water heat exchangers). Criterion 33, Reactor Coolant Makeup, may be less important for some reactor types than it is for water-cooled reactors. Residual heat removal is undoubtedly important for all reactor types, although the actuation time may not be critical for some designs.

Criteria 35 through 43, which address emergency core cooling and containment heat removal and cleanup, are probably not applicable to all reactor types, and certainly not as currently written. Criteria 44 through 46, which concern heat transfer to the ultimate heat sink, are appropriate for all reactors, but timing may not be critical for some designs.

The criteria in Group V, Reactor Containment, would not be applicable, as worded, to any reactor type for which a leak-tight containment is not required. Equivalent requirements on controlling or limiting release of fission products to the environment under accident conditions would seem necessary for all reactor types.

Criteria in Group VI, Fuel and Reactivity Control, are applicable to all reactor types.

9. COMMENTS ON RISK-INFORMING APPENDIX B

Appendix B to 10 CFR Part 50 contains the quality assurance criteria for nuclear power plants and fuel reprocessing plants. Attachment 4 to this report presents the complete text of Appendix B, which consists of an introduction and the following 18 criteria:

I	Organization
II	Quality Assurance Program
III	Design Control
IV	Procurement Document Control
V	Instructions, Procedures, and Drawings
VI	Document Control
VII	Control of Purchased Material, Equipment, and Services

VIII	Identification and Control of Materials, Parts, and Components
IX	Control of Special Processes
X	Inspection
XI	Test Control
XII	Control of Measuring and Test Equipment
XIII	Handling, Storage, and Shipping
XIV	Inspection, Test, and Operating Status
XV	Nonconforming Materials, Parts, or Components
XVI	Corrective Action
XVII	Quality Assurance Records
XVIII	Audits

Although Appendix B has often been identified as a source of unnecessary regulatory burden, critical evaluations have almost always concluded that Appendix B is structured to provide appropriate flexibility, and that the burden is created by the implementation of Appendix B through industry consensus standards and licensees' QA programs and the NRC's enforcement of those programs.

Appendix B defines quality assurance as "... all those planned and systematic actions necessary to provide adequate confidence that a structure, system, or component will perform satisfactorily in service." The scope of Appendix B, as established in its introduction, is "...all activities affecting the safety-related functions ... [of] ... structures, systems, and components that prevent or mitigate the consequences of postulated accidents that could cause undue risk to the health and safety of the public." The implied safety standard, thus, is "no undue risk" or "adequate protection," rather than risk metrics that are consistent with safety goals.

The introduction to Appendix B, Criterion I, and Criterion III use the term "safety-related." The risk focus could be improved by replacing "safety-related" with the term "important to risk," as previously discussed in regard to risk-informing the GDC. Criterion II, Quality Assurance Program, provides that "The applicant shall identify the structures, systems, and components to be covered by the quality assurance program ... " and "The quality assurance program shall provide control over activities affecting the quality of the identified structures, systems, and components, *to an extent consistent with their importance to safety*" (emphasis added). The italicized phrase appears to provide the flexibility needed to avoid committing resources to activities or hardware that is not safety-significant. Indeed, the industry and the NRC staff have generally agreed that Appendix B provides the flexibility to allow graded application of QA requirements on the basis of safety-significance of the activity or component. The NRC staff's position on graded QA programs is documented in Regulatory Guide 1.176 [14].

10. REFERENCES

1. SECY-98-300, "Options for Risk-Informed Revisions to 10 CFR Part 50, "Domestic Licensing of Production and Utilization Facilities," December 23, 1998.

2. Memorandum from Harold L. Price, Director of Regulation, to Chairman Seaborg, Commissioner Palfrey, Commissioner Ramey, and Commissioner Tape, "Proposed Design Criteria for Nuclear Power Plant Construction Permits," November 12, 1965.

3. SECY-86-284, "Progress Report on Review of Existing Regulatory Requirements," September 25, 1986.

4. SECY-91-224, "Elimination of Requirements Marginal to Safety," July 29, 1991.

5. SECY-92-263, "Staff Plans for Elimination of Requirements Marginal to Safety," July 24, 1992.

6. SECY-94-090, "Institutionalization of Continuing Program for Regulatory Improvement," May 18, 1994.

7. Frank Gillespie, et al., U.S. Nuclear Regulatory Commission, "Regulatory Review Group Report," August 1993.

8. Memorandum from Annette Vietti-Cook, Secretary, to William D. Travers, Executive Director for Operations, "Staff Requirements—SECY-98-300—Options for Risk-Informed Revisions to 10 CFR Part 50— "Domestic Licensing of Production and Utilization Facilities'," June 8, 1999.

9. SECY-99-256, "Rulemaking Plan for Risk-Informing Special Treatment Requirements," October 29, 1999.

10. SECY-00-0194, "Risk-Informing Special Treatment Requirements," September 7, 2000.

11. SECY-99-264, "Proposed Staff Plan for Risk-Informing Technical Requirements in 10 CFR Part 50," November 8, 1999.

12. SECY-00-0086, "Status Report on Risk-Informing the Technical Requirements of 10 CFR Part 50 (Option 3)," April 12, 2000.

13. SECY-00-0198, "Status Report on Study of Risk-Informed Changes to the Technical Requirements of 10 CFR Part 50 (Option 3) and Recommendations On Risk-Informed Changes to 10 CFR 50.44 (Combustible Gas Control)," September 14, 2000.

14. Regulatory Guide 1.176, "An Approach for Plant-Specific, Risk-Informed Decisionmaking: Graded Quality Assurance," August 1998.

APPENDIX A TO PART 50
GENERAL DESIGN CRITERIA FOR NUCLEAR POWER PLANTS

Introduction

Pursuant to the provisions of §50.34, an application for a construction permit must include the principal design criteria for a proposed facility. The principal design criteria establish the necessary design, fabrication, construction, testing, and performance requirements for structures, systems, and components important to safety; that is, structures, systems, and components that provide reasonable assurance that the facility can be operated without undue risk to the health and safety of the public.

These General Design Criteria establish minimum requirements for the principal design criteria for water-cooled nuclear power plants similar in design and location to plants for which construction permits have been issued by the Commission. The General Design Criteria are also considered to be generally applicable to other types of nuclear power units and are intended to provide guidance in establishing the principal design criteria for such other units.

The development of these General Design Criteria is not yet complete. For example, some of the definitions need further amplification. Also, some of the specific design requirements for structures, systems, and components important to safety have not as yet been suitably defined. Their omission does not relieve any applicant from considering these matters in the design of a specific facility and satisfying the necessary safety requirements. These matters include:

(1) Consideration of the need to design against single failures of passive components in fluid systems important to safety. (See Definition of Single Failure.)

(2) Consideration of redundancy and diversity requirements for fluid systems important to safety. A "system" could consist of a number of subsystems each of which is separately capable of performing the specified system safety function. The minimum acceptable redundancy and diversity of subsystems and components within a subsystem, and the required interconnection and independence of the subsystems have not yet been developed or defined. (See Criteria 34, 35, 38, 41, and 44.)

(3) Consideration of the type, size, and orientation of possible breaks in components of the reactor coolant pressure boundary in determining design requirements to suitably protect against postulated loss-of-coolant accidents. (See Definition of Loss of Coolant Accidents.)

(4) Consideration of the possibility of systematic, nonrandom, concurrent failures of redundant elements in the design of protection systems and reactivity control systems. (See Criteria 22, 24, 26, and 29.)

It is expected that the criteria will be augmented and changed from time to time as important new requirements for these and other features are developed.

There will be some water-cooled nuclear power plants for which the General Design Criteria are not sufficient and for which additional criteria must be identified and satisfied in the interest of public safety. In particular, it is expected that additional or different criteria will be needed to take into account unusual sites and environmental conditions, and for water-cooled nuclear power units of advanced design. Also, there may be water-cooled nuclear power units for which fulfillment of some of the General Design Criteria may not be necessary or appropriate. For plants such as these, departures from the General Design Criteria must be identified and justified.

Definitions and Explanations

Nuclear power unit. A nuclear power unit means a nuclear power reactor and associated equipment necessary for electric power generation and includes those structures, systems, and components required to provide reasonable assurance the facility can be operated without undue risk to the health and safety of the public.

Loss of coolant accidents. Loss of coolant accidents mean those postulated accidents that result from the loss of reactor coolant at a rate in excess of the capability of the reactor coolant makeup system from breaks in the reactor coolant pressure boundary, up to and including a break equivalent in size to the double-ended rupture of the largest pipe of the reactor coolant system.(1)

Single failure. A single failure means an occurrence which results in the loss of capability of a component to perform its intended safety functions. Multiple failures resulting from a single occurrence are considered to be a single failure. Fluid and electric systems are considered to be designed against an assumed single failure if neither (1) a single failure of any active component (assuming passive components function properly) nor (2) a single failure of a passive component (assuming active components function properly), results in a loss of the capability of the system to perform its safety functions.(2)

Anticipated operational occurrences. Anticipated operational occurrences mean those conditions of normal operation which are expected to occur one or more times during the life of the nuclear power unit and include but are not limited to loss of power to all recirculation pumps, tripping of the turbine generator set, isolation of the main condenser, and loss of all offsite power.

Criteria

I. Overall Requirements

Criterion 1 -- Quality standards and records. Structures, systems, and components important to safety shall be designed, fabricated, erected, and tested to quality standards commensurate with the importance of the safety functions to be performed. Where generally recognized codes and standards are used, they shall be identified and evaluated to determine their applicability, adequacy, and sufficiency and shall be supplemented or modified as necessary to assure a quality product in keeping with the required safety function. A quality assurance program shall be established and implemented in order to provide adequate assurance that these structures, systems, and components will satisfactorily perform their safety functions. Appropriate records of the design, fabrication, erection, and testing of structures, systems, and components important to safety shall be maintained by or under the control of the nuclear power unit licensee throughout the life of the unit.

Criterion 2 -- Design bases for protection against natural phenomena. Structures, systems, and components important to safety shall be designed to withstand the effects of natural phenomena such as earthquakes, tornadoes, hurricanes, floods, tsunami, and seiches without loss of capability to perform their safety functions. The design bases for these structures, systems, and components shall reflect: (1) Appropriate consideration of the most severe of the natural phenomena that have been historically reported for the site and surrounding area, with sufficient margin for the limited accuracy, quantity, and period of time in which the historical data have been accumulated, (2) appropriate combinations of the effects of normal and accident conditions with the effects of the natural phenomena and (3) the importance of the safety functions to be performed.

Criterion 3 -- Fire protection. Structures, systems, and components important to safety shall be designed and located to minimize, consistent with other safety requirements, the probability and effect of fires and explosions. Noncombustible and heat resistant materials shall be used wherever practical throughout the unit, particularly in locations such as the containment and control room. Fire detection and fighting systems of appropriate capacity and capability shall be provided and designed to minimize the adverse effects of fires on structures, systems, and components important to safety. Firefighting systems shall be designed to assure that their rupture or inadvertent operation does not significantly impair the safety capability of these structures, systems, and components.

Criterion 4 -- Environmental and dynamic effects design bases. Structures, systems, and components important to safety shall be designed to accommodate the effects of and to be compatible with the environmental conditions associated with normal operation, maintenance, testing, and postulated accidents, including loss-of-coolant accidents. These structures, systems, and components shall be appropriately protected against dynamic effects, including the effects of missiles, pipe whipping, and discharging fluids, that may result from equipment failures and from events and conditions outside the nuclear power unit. However, dynamic effects associated with postulated pipe ruptures in nuclear power units may be excluded from the design basis when analyses reviewed and approved by the Commission demonstrate that the probability of fluid system piping rupture is extremely low under conditions consistent with the design basis for the piping.

Criterion 5 -- Sharing of structures, systems, and components. Structures, systems, and components important to safety shall not be shared among nuclear power units unless it can be shown that such sharing will not significantly impair their ability to perform their safety functions, including, in the event of an accident in one unit, an orderly shutdown and cooldown of the remaining units.

II. Protection by Multiple Fission Product Barriers

Criterion 10 -- Reactor design. The reactor core and associated coolant, control, and protection systems shall be designed with appropriate margin to assure that specified acceptable fuel design limits are not exceeded during any condition of normal operation, including the effects of anticipated operational occurrences.

Criterion 11 -- Reactor inherent protection. The reactor core and associated coolant systems shall be designed so that in the power operating range the net effect of the prompt inherent nuclear feedback characteristics tends to compensate for a rapid increase in reactivity.

Criterion 12 -- Suppression of reactor power oscillations. The reactor core and associated coolant, control, and protection systems shall be designed to assure that power oscillations which can result in conditions exceeding specified acceptable fuel design limits are not possible or can be reliably and readily detected and suppressed.

Criterion 13 -- Instrumentation and control. Instrumentation shall be provided to monitor variables and systems over their anticipated ranges for normal operation, for anticipated operational occurrences, and for accident conditions as appropriate to assure adequate safety, including those variables and systems that can affect the fission process, the integrity of the reactor core, the reactor coolant pressure boundary, and the containment and its associated systems. Appropriate controls shall be provided to maintain these variables and systems within prescribed operating ranges.

Criterion 14 -- Reactor coolant pressure boundary. The reactor coolant pressure boundary shall be designed, fabricated, erected, and tested so as to have an extremely low probability of abnormal leakage, of rapidly propagating failure, and of gross rupture.

Criterion 15 -- Reactor coolant system design. The reactor coolant system and associated auxiliary, control, and protection systems shall be designed with sufficient margin to assure that the design conditions of the reactor coolant pressure boundary are not exceeded during any condition of normal operation, including anticipated operational occurrences.

Criterion 16 -- Containment design. Reactor containment and associated systems shall be provided to establish an essentially leak-tight barrier against the uncontrolled release of radioactivity to the environment and to assure that the containment design conditions important to safety are not exceeded for as long as postulated accident conditions require.

Criterion 17 -- Electric power systems. An onsite electric power system and an offsite electric power system shall be provided to permit functioning of structures, systems, and components important to safety. The safety function for each system (assuming the other system is not functioning) shall be to provide sufficient capacity and capability to assure that (1) specified acceptable fuel design limits and design conditions of the reactor coolant pressure boundary are not exceeded as a result of anticipated operational occurrences and (2) the core is cooled and containment integrity and other vital functions are maintained in the event of postulated accidents.

The onsite electric power supplies, including the batteries, and the onsite electric distribution system, shall have sufficient independence, redundancy, and testability to perform their safety functions assuming a single failure.

Electric power from the transmission network to the onsite electric distribution system shall be supplied by two physically independent circuits (not necessarily on separate rights of way) designed and located so as to minimize to the extent practical the likelihood of their simultaneous failure under operating and postulated accident and environmental conditions. A switchyard common to both circuits is acceptable. Each of these circuits shall be designed to be available in sufficient time following a loss of all onsite alternating current power supplies and the other offsite electric power circuit, to assure that specified acceptable fuel design limits and design conditions of the reactor coolant pressure boundary are not exceeded. One of these circuits shall be designed to be available within a few seconds following a loss-of-coolant accident to assure that core cooling, containment integrity, and other vital safety functions are maintained.

Provisions shall be included to minimize the probability of losing electric power from any of the remaining supplies as a result of, or coincident with, the loss of power generated by the nuclear power unit, the loss of power from the transmission network, or the loss of power from the onsite electric power supplies.

Criterion 18 -- Inspection and testing of electric power systems. Electric power systems important to safety shall be designed to permit appropriate periodic inspection and testing of important areas and features, such as wiring, insulation, connections, and switchboards, to assess the continuity of the systems and the condition of their components. The systems shall be designed with a capability to test periodically (1) the operability and functional performance of the components of the systems, such as onsite power sources, relays, switches, and buses, and (2) the operability of the systems as a whole and, under conditions as close to design as practical, the full operation sequence that brings the systems into operation, including operation of applicable portions of the protection system, and the transfer of power among the nuclear power unit, the offsite power system, and the onsite power system.

Criterion 19 -- Control room. A control room shall be provided from which actions can be taken to operate the nuclear power unit safely under normal conditions and to maintain it in a safe condition under accident conditions, including loss-of-coolant accidents. Adequate radiation protection shall be provided to permit access and occupancy of the control room under accident conditions without personnel receiving radiation exposures in excess of 5 rem whole body, or its equivalent to any part of the body, for the duration of the accident.

Equipment at appropriate locations outside the control room shall be provided (1) with a design capability for prompt hot shutdown of the reactor, including necessary instrumentation and controls to maintain the unit in a safe condition during hot shutdown, and (2) with a potential capability for subsequent cold shutdown of the reactor through the use of suitable procedures.

Applicants for and holders of construction permits and operating licenses under this part who apply on or after January 10, 1997, applicants for design certifications under part 52 of this chapter who apply on or after January 10, 1997, applicants for and holders of combined licenses under part 52 of this chapter who do not reference a standard design certification, or holders of operating licenses using an alternative source term under section 50.67 shall meet the requirement of this criterion, except that with regard to control room access and occupancy, adequate radiation protection shall be provided to ensure that radiation exposures shall not exceed 0.05 Sv (5 rem) total effective dose equivalent (TEDE) as defined in section 50.2 for the duration of the accident.

III. Protection and Reactivity Control Systems

Criterion 20 -- Protection system functions. The protection system shall be designed (1) to initiate automatically the operation of appropriate systems including the reactivity control systems, to assure that specified acceptable fuel design limits are not exceeded as a result of anticipated operational occurrences and (2) to sense accident conditions and to initiate the operation of systems and components important to safety.

Criterion 21 -- Protection system reliability and testability. The protection system shall be designed for high functional reliability and inservice testability commensurate with the safety functions to be performed. Redundancy and independence designed into the protection system shall be sufficient to assure that (1) no single failure results in loss of the protection function and (2) removal from service of any component or channel does not result in loss of the required minimum redundancy unless the acceptable reliability of operation of the protection system can be otherwise demonstrated. The protection system shall be designed to permit periodic testing of its functioning when the reactor is in operation, including a capability to test channels independently to determine failures and losses of redundancy that may have occurred.

Criterion 22 -- Protection system independence. The protection system shall be designed to assure that the effects of natural phenomena, and of normal operating, maintenance, testing, and postulated accident conditions on redundant channels do not result in loss of the protection function, or shall be demonstrated to be acceptable on some other defined basis. Design techniques, such as functional diversity or diversity in component design and principles of operation, shall be used to the extent practical to prevent loss of the protection function.

Criterion 23 -- Protection system failure modes. The protection system shall be designed to fail into a safe state or into a state demonstrated to be acceptable on some other defined basis if conditions such as disconnection of the system, loss of energy (e.g., electric power, instrument air), or postulated adverse environments (e.g., extreme heat or cold, fire, pressure, steam, water, and radiation) are experienced.

Criterion 24 -- Separation of protection and control systems. The protection system shall be separated from control systems to the extent that failure of any single control system component or channel, or failure or removal from service of any single protection system component or channel which is common to the control and protection systems leaves intact a system satisfying all reliability, redundancy, and independence requirements of the protection system. Interconnection of the protection and control systems shall be limited so as to assure that safety is not significantly impaired.

Criterion 25 -- Protection system requirements for reactivity control malfunctions. The protection system shall be designed to assure that specified acceptable fuel design limits are not exceeded for any single malfunction of the reactivity control systems, such as accidental withdrawal (not ejection or dropout) of control rods.

Criterion 26 -- Reactivity control system redundancy and capability. Two independent reactivity control systems of different design principles shall be provided. One of the systems shall use control rods, preferably including a positive means for inserting the rods, and shall be capable of reliably controlling reactivity changes to assure that under conditions of normal operation, including anticipated operational occurrences, and with appropriate margin for malfunctions such as stuck rods, specified acceptable fuel design limits are not exceeded. The second reactivity control system shall be capable of reliably controlling the rate of reactivity changes resulting from planned, normal power changes (including xenon burnout) to assure acceptable fuel design limits are not exceeded. One of the systems shall be capable of holding the reactor core subcritical under cold conditions.

Criterion 27 -- Combined reactivity control systems capability. The reactivity control systems shall be designed to have a combined capability, in conjunction with poison addition by the emergency core cooling system, of reliably controlling reactivity changes to assure that under postulated accident conditions and with appropriate margin for stuck rods the capability to cool the core is maintained.

Criterion 28 -- Reactivity limits. The reactivity control systems shall be designed with appropriate limits on the potential amount and rate of reactivity increase to assure that the effects of postulated reactivity accidents can neither (1) result in damage to the reactor coolant pressure boundary greater than limited local yielding nor (2) sufficiently disturb the core, its support structures or other reactor pressure vessel internals to impair significantly the capability to cool the core. These postulated reactivity accidents shall include consideration of rod ejection (unless prevented by positive means), rod dropout, steam line rupture, changes in reactor coolant temperature and pressure, and cold water addition.

Criterion 29 -- Protection against anticipated operational occurrences. The protection and reactivity control systems shall be designed to assure an extremely high probability of accomplishing their safety functions in the event of anticipated operational occurrences.

IV. Fluid Systems

Criterion 30 -- Quality of reactor coolant pressure boundary. Components which are part of the reactor coolant pressure boundary shall be designed, fabricated, erected, and tested to the highest quality standards practical. Means shall be provided for detecting and, to the extent practical, identifying the location of the source of reactor coolant leakage.

Criterion 31 -- Fracture prevention of reactor coolant pressure boundary. The reactor coolant pressure boundary shall be designed with sufficient margin to assure that when stressed under operating, maintenance, testing, and postulated accident conditions (1) the boundary behaves in a nonbrittle manner and (2) the probability of rapidly propagating fracture is minimized. The design shall reflect consideration of service temperatures and other conditions of the boundary material under operating, maintenance, testing, and postulated accident conditions and the uncertainties in determining (1) material properties, (2) the effects of irradiation on material properties, (3) residual, steady state and transient stresses, and (4) size of flaws.

Criterion 32 -- Inspection of reactor coolant pressure boundary. Components which are part of the reactor coolant pressure boundary shall be designed to permit (1) periodic inspection and testing of important areas and features to assess their structural and leaktight integrity, and (2) an appropriate material surveillance program for the reactor pressure vessel.

Criterion 33 -- Reactor coolant makeup. A system to supply reactor coolant makeup for protection against small breaks in the reactor coolant pressure boundary shall be provided. The system safety function shall be to assure that specified acceptable fuel design limits are not exceeded as a result of reactor coolant loss due to leakage from the reactor coolant pressure boundary and rupture of small piping or other small components which are part of the boundary. The system shall be designed to assure that for onsite electric power system operation (assuming offsite power is not available) and for offsite electric power system operation (assuming onsite power is not available) the system safety function can be accomplished using the piping, pumps, and valves used to maintain coolant inventory during normal reactor operation.

Criterion 34 -- Residual heat removal. A system to remove residual heat shall be provided. The system safety function shall be to transfer fission product decay heat and other residual heat from the reactor core at a rate such that specified acceptable fuel design limits and the design conditions of the reactor coolant pressure boundary are not exceeded.

Suitable redundancy in components and features, and suitable interconnections, leak detection, and isolation capabilities shall be provided to assure that for onsite electric power system operation (assuming offsite power is not available) and for offsite electric power system operation (assuming onsite power is not available) the system safety function can be accomplished, assuming a single failure.

Criterion 35 -- Emergency core cooling. A system to provide abundant emergency core cooling shall be provided. The system safety function shall be to transfer heat from the reactor core following any loss of reactor coolant at a rate such that (1) fuel and clad damage that could interfere with continued effective core cooling is prevented and (2) clad metal-water reaction is limited to negligible amounts.

Suitable redundancy in components and features, and suitable interconnections, leak detection, isolation, and containment capabilities shall be provided to assure that for onsite electric power system operation (assuming offsite power is not available) and for offsite electric power system operation (assuming onsite power is not available) the system safety function can be accomplished, assuming a single failure.

Criterion 36 -- Inspection of emergency core cooling system. The emergency core cooling system shall be designed to permit appropriate periodic inspection of important components, such as spray rings in the reactor pressure vessel, water injection nozzles, and piping, to assure the integrity and capability of the system.

Criterion 37 -- Testing of emergency core cooling system. The emergency core cooling system shall be designed to permit appropriate periodic pressure and functional testing to assure (1) the structural and leaktight integrity of its components, (2) the operability and performance of the active components of the system, and (3) the operability of the system as a whole and, under conditions as close to design as practical, the performance of the full operational sequence that brings the system into operation, including operation of applicable portions of the protection system, the transfer between normal and emergency power sources, and the operation of the associated cooling water system.

Criterion 38 -- Containment heat removal. A system to remove heat from the reactor containment shall be provided. The system safety function shall be to reduce rapidly, consistent with the functioning of other associated systems, the containment pressure and temperature following any loss-of-coolant accident and maintain them at acceptably low levels.

Suitable redundancy in components and features, and suitable interconnections, leak detection, isolation, and containment capabilities shall be provided to assure that for onsite electric power system operation (assuming offsite power is not available) and for offsite electric power system operation (assuming onsite power is not available) the system safety function can be accomplished, assuming a single failure.

Criterion 39 -- Inspection of containment heat removal system. The containment heat removal system shall be designed to permit appropriate periodic inspection of important components, such as the torus, sumps, spray nozzles, and piping to assure the integrity and capability of the system.

Criterion 40 -- Testing of containment heat removal system. The containment heat removal system shall be designed to permit appropriate periodic pressure and functional testing to assure (1) the structural and leaktight integrity of its components, (2) the operability and performance of the active components of the system, and (3) the operability of the system as a whole, and under conditions as close to the design as practical the performance of the full operational sequence that brings the system into operation, including operation of applicable portions of the protection system, the transfer between normal and emergency power sources, and the operation of the associated cooling water system.

Criterion 41 -- Containment atmosphere cleanup. Systems to control fission products, hydrogen, oxygen, and other substances which may be released into the reactor containment shall be provided as necessary to reduce, consistent with the functioning of other associated systems, the concentration and quality of fission products released to the environment following postulated accidents, and to control the concentration of hydrogen or oxygen and other substances in the containment atmosphere following postulated accidents to assure that containment integrity is maintained.

Each system shall have suitable redundancy in components and features, and suitable interconnections, leak detection, isolation, and containment capabilities to assure that for onsite electric power system operation (assuming offsite power is not available) and for offsite electric power system operation (assuming onsite power is not available) its safety function can be accomplished, assuming a single failure.

Criterion 42 -- Inspection of containment atmosphere cleanup systems. The containment atmosphere cleanup systems shall be designed to permit appropriate periodic inspection of important components, such as filter frames, ducts, and piping to assure the integrity and capability of the systems.

Criterion 43 -- Testing of containment atmosphere cleanup systems. The containment atmosphere cleanup systems shall be designed to permit appropriate periodic pressure and functional testing to assure (1) the structural and leaktight integrity of its components, (2) the operability and performance of the active components of the systems such as fans, filters, dampers, pumps, and valves and (3) the operability of the systems as a whole and, under conditions as close to design as practical, the performance of the full operational sequence that brings the systems into operation, including operation of applicable portions of the protection system, the transfer between normal and emergency power sources, and the operation of associated systems.

Criterion 44 -- Cooling water. A system to transfer heat from structures, systems, and components important to safety, to an ultimate heat sink shall be provided. The system safety function shall be to transfer the combined heat load of these structures, systems, and components under normal operating and accident conditions.

Suitable redundancy in components and features, and suitable interconnections, leak detection, and isolation capabilities shall be provided to assure that for onsite electric power system operation (assuming offsite power is not available) and for offsite electric power system operation (assuming onsite power is not available) the system safety function can be accomplished, assuming a single failure.

Criterion 45 -- Inspection of cooling water system. The cooling water system shall be designed to permit appropriate periodic inspection of important components, such as heat exchangers and piping, to assure the integrity and capability of the system.

Criterion 46 -- Testing of cooling water system. The cooling water system shall be designed to permit appropriate periodic pressure and functional testing to assure (1) the structural and leaktight integrity of its components, (2) the operability and the performance of the active components of the system, and (3) the operability of the system as a whole and, under conditions as close to design as practical, the performance of the full operational sequence that brings the system into operation for reactor shutdown and for loss-of-coolant accidents, including operation of applicable portions of the protection system and the transfer between normal and emergency power sources.

V. Reactor Containment

Criterion 50 -- Containment design basis. The reactor containment structure, including access openings, penetrations, and the containment heat removal system shall be designed so that the containment structure and its internal compartments can accommodate, without exceeding the design leakage rate and with sufficient margin, the calculated pressure and temperature conditions resulting from any loss-of-coolant accident. This margin shall reflect consideration of (1) the effects of potential energy sources which have not been included in the determination of the peak conditions, such as energy in steam generators and as required by §50.44 energy from metal-water and other chemical reactions that may result from degradation but not total failure of emergency core cooling functioning, (2) the limited experience and experimental data available for defining accident phenomena and containment responses, and (3) the conservatism of the calculational model and input parameters.

Criterion 51 -- Fracture prevention of containment pressure boundary. The reactor containment boundary shall be designed with sufficient margin to assure that under operating, maintenance, testing, and postulated accident conditions (1) its ferritic materials behave in a nonbrittle manner and (2) the probability of rapidly propagating fracture is minimized. The design shall reflect consideration of service temperatures and other conditions of the containment boundary material during operation, maintenance, testing, and postulated accident conditions, and the uncertainties in determining (1) material properties, (2) residual, steady state, and transient stresses, and (3) size of flaws.

Criterion 52 -- Capability for containment leakage rate testing. The reactor containment and other equipment which may be subjected to containment test conditions shall be designed so that periodic integrated leakage rate testing can be conducted at containment design pressure.

Criterion 53 -- Provisions for containment testing and inspection. The reactor containment shall be designed to permit (1) appropriate periodic inspection of all important areas, such as penetrations, (2) an appropriate surveillance program, and (3) periodic testing at containment design pressure of the leaktightness of penetrations which have resilient seals and expansion bellows.

Criterion 54 -- Piping systems penetrating containment. Piping systems penetrating primary reactor containment shall be provided with leak detection, isolation, and containment capabilities having redundancy, reliability, and performance capabilities which reflect the importance to safety of isolating these piping systems. Such piping systems shall be designed with a capability to test periodically the operability of the isolation valves and associated apparatus and to determine if valve leakage is within acceptable limits.

Criterion 55 -- Reactor coolant pressure boundary penetrating containment. Each line that is part of the reactor coolant pressure boundary and that penetrates primary reactor containment shall be provided with containment isolation valves as follows, unless it can be demonstrated that the containment isolation provisions for a specific class of lines, such as instrument lines, are acceptable on some other defined basis:

(1) One locked closed isolation valve inside and one locked closed isolation valve outside containment; or

(2) One automatic isolation valve inside and one locked closed isolation valve outside containment; or

(3) One locked closed isolation valve inside and one automatic isolation valve outside containment. A simple check valve may not be used as the automatic isolation valve outside containment; or

(4) One automatic isolation valve inside and one automatic isolation valve outside containment. A simple check valve may not be used as the automatic isolation valve outside containment.

Isolation valves outside containment shall be located as close to containment as practical and upon loss of actuating power, automatic isolation valves shall be designed to take the position that provides greater safety.

Other appropriate requirements to minimize the probability or consequences of an accidental rupture of these lines or of lines connected to them shall be provided as necessary to assure adequate safety. Determination of the appropriateness of these requirements, such as higher quality in design, fabrication, and testing, additional provisions for inservice inspection, protection against more severe natural phenomena, and additional isolation valves and containment, shall include consideration of the population density, use characteristics, and physical characteristics of the site environs.

Criterion 56 -- Primary containment isolation. Each line that connects directly to the containment atmosphere and penetrates primary reactor containment shall be provided with containment isolation valves as follows, unless it can be demonstrated that the containment isolation provisions for a specific class of lines, such as instrument lines, are acceptable on some other defined basis:

(1) One locked closed isolation valve inside and one locked closed isolation valve outside containment; or

(2) One automatic isolation valve inside and one locked closed isolation valve outside containment; or

(3) One locked closed isolation valve inside and one automatic isolation valve outside containment. A simple check valve may not be used as the automatic isolation valve outside containment; or

(4) One automatic isolation valve inside and one automatic isolation valve outside containment. A simple check valve may not be used as the automatic isolation valve outside containment.

Isolation valves outside containment shall be located as close to the containment as practical and upon loss of actuating power, automatic isolation valves shall be designed to take the position that provides greater safety.

Criterion 57 -- Closed system isolation valves. Each line that penetrates primary reactor containment and is neither part of the reactor coolant pressure boundary nor connected directly to the containment atmosphere shall have at least one containment isolation valve which shall be either automatic, or locked closed, or capable of remote manual operation. This valve shall be outside containment and located as close to the containment as practical. A simple check valve may not be used as the automatic isolation valve.

VI. Fuel and Radioactivity Control

Criterion 60 -- Control of releases of radioactive materials to the environment. The nuclear power unit design shall include means to control suitably the release of radioactive materials in gaseous and liquid effluents and to handle radioactive solid wastes produced during normal reactor operation, including anticipated operational occurrences. Sufficient holdup capacity shall be provided for retention of gaseous and liquid effluents containing radioactive materials, particularly where unfavorable site environmental conditions can be expected to impose unusual operational limitations upon the release of such effluents to the environment.

Criterion 61 -- Fuel storage and handling and radioactivity control. The fuel storage and handling, radioactive waste, and other systems which may contain radioactivity shall be designed to assure adequate safety under normal and postulated accident conditions. These systems shall be designed (1) with a capability to permit appropriate periodic inspection and testing of components important to safety, (2) with suitable shielding for radiation protection, (3) with appropriate containment, confinement, and filtering systems, (4) with a residual heat removal capability having reliability and testability that reflects the importance to safety of decay heat and other residual heat removal, and (5) to prevent significant reduction in fuel storage coolant inventory under accident conditions.

Criterion 62 -- Prevention of criticality in fuel storage and handling. Criticality in the fuel storage and handling system shall be prevented by physical systems or processes, preferably by use of geometrically safe configurations.

Criterion 63 -- Monitoring fuel and waste storage. Appropriate systems shall be provided in fuel storage and radioactive waste systems and associated handling areas (1) to detect conditions that may result in loss of residual heat removal capability and excessive radiation levels and (2) to initiate appropriate safety actions.

Criterion 64 -- Monitoring radioactivity releases. Means shall be provided for monitoring the reactor containment atmosphere, spaces containing components for recirculation of loss-of-coolant accident fluids, effluent discharge paths, and the plant environs for radioactivity that may be released from normal operations, including anticipated operational occurrences, and from postulated accidents.

[36 FR 3256, Feb. 20, 1971, as amended at 36 FR 12733, July 7, 1971; 41 FR 6258, Feb. 12, 1976; 43 FR 50163, Oct. 27, 1978; 51 FR 12505, Apr. 11, 1986; 52 FR 41294, Oct. 27, 1987]

Notes:

1 Further details relating to the type, size, and orientation of postulated breaks in specific components of the reactor coolant pressure boundary are under development.

2 Single failures of passive components in electric systems should be assumed in designing against a single failure. The conditions under which a single failure of a passive component in a fluid system should be considered in designing the system against a single failure are under development.

PROPOSED DESIGN CRITERIA FOR
NUCLEAR POWER PLANT CONSTRUCTION PERMITS - 1965

A November 12, 1965 memorandum from Harold Price, Director of Regulation, to the Chairman and Commissioners of the Atomic Energy Commission transmitted Proposed General Design Criteria for Nuclear Power Plant Construction Permits. The criteria were sent to the Commission for its consideration, and were also to be released for public comment and interim guidance to the industry. The memo notes that the ACRS had approved a specific change to one of the criteria.

The following text is a verbatim copy of the 1965 draft criteria. The available copy is illegible in a few places, and in Criteria 4 and 8, there are two or three words that could not be deciphered. Rather than guess at the missing words, the parenthetical note (illegible) has been inserted.

FACILITY

Criterion 1: those features of reactor facilities which are essential to the prevention of accidents or to the mitigation of their consequences must be designed, fabricated and erected to:

(a) Quality standards that reflect the importance of the safety function to be performed. It should be recognized, in this respect, that design codes commonly used for nonnuclear applications may not be adequate.

(b) Performance standards that will enable the facility to withstand, without loss of the capability to protect the public, the additional forces imposed by the most severe earthquakes, flooding conditions, winds, ice, and other natural phenomena anticipated at the proposed site.

Criterion 2: Provisions must be included to limit the extent and consequences of credible chemical reactions that could cause or materially augment the release of significant amounts of fission products from the facility.

Criterion 3: Protection must be provided against possibilities for damage of the safeguarding features of the facility by missiles generated through equipment failures inside the containment.

REACTOR

Criterion 4: The reactor must be designed to accommodate, without fuel failure or primary system damage (illegible) steady state norm that might be (illegible) anticipated transient events such as tripping of the turbine-generator and loss of power to the reactor recirculation system pumps.

Criterion 5: The reactor must be designed so that power or process variable oscillations or transients that could cause fuel failure or primary system damage are not possible or can be readily suppressed.

Criterion 6: Clad fuel must be designed to accommodate throughout its design lifetime all normal and abnormal modes of anticipated reactor operation, including the design overpower condition, without experiencing significant cladding failures. Unclad or vented fuels must be designed with the similar objective of providing control over fission products. For unclad and vented solid fuels, normal and abnormal modes of anticipated reactor operation must be achieved without exceeding design release rates of fission products from the fuel over core lifetime.

Criterion 7: The maximum reactivity worth of control rods or elements and the rates with which reactivity can be inserted must be held to values such that no single credible mechanical or electrical control system malfunction could cause a reactivity transient capable of damaging the primary system or cause significant fuel failure.

Criterion 8: (illegible) capability must be provided to make and hold the core subcritical (illegible) any credible operating condition with any one control element at its position of highest reactivity.

Criterion 9: Backup reactivity shutdown capability must be provided that is independent of normal reactivity control provisions. This system must have the capability to shut down the reactor from any operating condition.

Criterion 10: Heat removal systems must be provided which are capable of accommodating core decay heat under all anticipated abnormal and credible accident conditions, such as isolation from the rain condenser and complete or partial loss of primary coolant from the reactor.

Criterion 11: Components of the primary coolant and containment systems must be designed and operated so that no substantial pressure or thermal stress will be imposed on the structural materials unless the temperatures are well above the nil-ductility temperatures. For ferritic materials of the coolant envelope and the containment, minimum temperatures are NDT + 60°F and NDT + 30°F, respectively.

Criterion 12: Capability for control rod insertion under abnormal conditions must be provided.

Criterion 13: The reactor facility must be provided with a control room from which all actions can be controlled or monitored as necessary to maintain safe operational status of the plant at all times. The control room must be provided with adequate protection to permit occupancy under the conditions described in Criterion 17 below, and with the means to shut down the plant and maintain it in a safe condition if such accident were to be experienced.

Criterion 14: Means must be included in the control room to show the relative reactivity status of the reactor such as position indication of mechanical rods or concentrations of chemical poisons.

Criterion 15: A reliable reactor protection system must be provided to automatically initiate appropriate action to prevent safety limits from being exceeded. Capability must be provided for testing functional operability of the system and for determining that no component or circuit failure has occurred. For instruments and control systems in vital areas where the potential consequences of failure require redundancy, the redundant channels must be independent and must be capable of being tested to determine that they remain independent. Sufficient redundancy must be provided that failure or removal from service of a single component or channel will not inhibit necessary safety action when required. These criteria should, where applicable, be satisfied by the instrumentation associated with containment closure and isolation systems, afterheat removal and core cooling systems, systems to prevent cold-slug accidents, and other vital systems, as well as the reactor nuclear and process safety system.

Criterion 16: The vital instrumentation systems of Criterion 15 must be designed so that no credible combination of circumstances can interfere with the performance of a safety function when it is needed. In particular, the effect of influences common to redundant channels which are intended to be independent must not negate the operability of a safety system. The effects of gross disconnection of the system, loss of energy (electric power, instrument air), and adverse environment (heat from loss of instrument cooling, extreme cold, fire, steam, water, etc.) must cause the system to go into its safest state (fail-safe) or be demonstrably tolerable on some other basis.

ENGINEERED SAFEGUARDS

Criterion 17: The containment structure, including access openings and penetrations, must be designed and fabricated to accommodate or dissipate without failure the pressures and temperatures associated with the largest credible energy release including the effects of credible metal-water or other chemical reactions uninhibited by active quenching systems. If part of the primary coolant system is outside the primary reactor containment, appropriate safeguards [such as additional containment] must be provided for that part if necessary, to protect the health and safety of the public, in case of an accidental rupture in that part of the system. The appropriateness of safeguards such as isolation valves, additional containment, etc., will depend on environmental and population conditions surrounding the site. (Emphasis in original.)

Criterion 18: Provisions must be made for the removal of heat from within the containment structure as necessary to maintain the integrity of the structure under the conditions described in Criterion 17 above. If engineered safeguards are needed to prevent containment vessel failure due to heat released under such conditions, at least two independent systems must be provided, preferably of different principles. Backup equipment (e.g., water and power systems) to such engineered safeguards must also be redundant.

Criterion 19: The maximum integrated from the containment structure under the conditions described in Criterion 17 above must meet the site exposure criteria set forth in 10 CFR 100. The containment structure must be designed so that the containment can be leak tested at least to design pressure conditions after completion and installation of all penetrations, and the leakage rate measured over a suitable period to verify its conformance with required performance. The plant must be designed for later tests at suitable pressures.

Criterion 20: All containment structure penetrations subject to failure such as resilient seals and expansion bellows must be designed and constructed so that leak-tightness can be demonstrated at design pressure at any time throughout operating life of the reactor.

Criterion 21: Sufficient normal and emergency sources of electrical power must be provided to assure a capability for prompt shutdown and continued maintenance of the reactor facility in a safe condition under all credible circumstances.

Criterion 22: Valves and their associated apparatus that are essential to the containment function must be redundant and so arranged that no credible combination of circumstances can interfere with their necessary functioning. Such redundant valves and associated apparatus must be independent of each other. Capability must be provided for testing functional operability of these valves and associated equipment to determine that no failure has occurred and that leakage is within acceptable limits. Redundant valves and auxiliaries must be independent. Containment closure valves must be actuated by instrumentation, control circuits and energy sources which satisfy Criterion 15 and 16 above.

Criterion 23: In determining the suitability of a facility for a proposed site the acceptance of the inherent and engineered safety afforded by the systems, materials and components, and the associated engineered safeguards built into the facility, will depend on their demonstrated performance capability and reliability and the extent to which the operability of such systems, materials, components, and engineered safeguards can be tested and inspected during the life of the plant.

RADIOACTIVITY CONTROL

Criterion 24: All fuel storage and waste handling systems must be contained if necessary to prevent the accidental release of radioactivity in amounts which could affect the health and safety of the public.

Criterion 25: The fuel handling and storage facilities must be designed to prevent criticality and to maintain adequate shielding and cooling for spent fuel under all anticipated normal and abnormal conditions, and credible accident conditions. Variables upon which health and safety of the public depend must be monitored.

Criterion 26: Where unfavorable environmental conditions can be expected to require limitations upon the release of operational radioactive effluents to the environment, appropriate hold-up capacity must be provided for retention of gaseous, liquid, or solid effluents.

Criterion 27: The plant must be provided with systems capable of monitoring the release of radioactivity under accident conditions.

SOME THOUGHTS ON RISK INFORMING THE GENERAL DESIGN CRITERIA

Thomas S. Kress

March 2, 2001

The General Design Criteria (GDC), as expressed in Appendix A to 10CFR50, place constraints on how particular important-to-safety functions are to be achieved in the nuclear plant's design to meet the requirements (figures-of-merit) of the FSAR Chapter 15 Design Basis Accidents (DBA). They "establish minimum requirements for principal design criteria which establish necessary design, fabrication, construction, testing, and performance requirements for structures, systems, and components important to safety.

The GDC are presented in 6 categories: Overall, Protection by Multiple Fission Product Barriers, Protection and Reactivity Control Systems, Fluid Systems, Reactor Containment, and Fuel and Radioactivity Control. Within these 6 categories, there are a total of 55 separate GDC. Clearly, these GDC are intended to assure high confidence in the performance of certain functions:

1. Power control
   a. Shutting off the reactor power
   b. Control of undue power oscillations
   c. Prevention of rapid power increases

2. Keeping the core , RCS, and containment cool under normal operation and accident conditions.

3. Providing protection against fission product release
   - normal operation
   - accidents
   - control room habitability

4. Protecting against external events.

The GDC also introduce the following concepts with respect to the design of SSCs to accommodate:

- single failures,
- redundancy and diversity,
- a range of break sizes,
- common cause failure of protection and reactivity control systems,

- quality assurance, and
- monitoring, inspection, and testing.

In addition, it is clear that there are to be multiple barriers against fission product release, although this does not seem to be explicitly required.

The GDC are the embodiment of deterministic, prescriptive regulation with a classical defense-in-depth focus. As such, they would appear to be the antithesis of risk-informed regulation. The objective of this write-up, then, is to provide some personal thoughts on how the GDC can be reconstituted so that the desired functional objectives can still be achieved in a way that the GDC are risk-informed as well as being generalized to apply to any reactor type (not just LWRs).

To achieve these broad objectives (particularly for generalized application) would appear to me to require wholesale revision of the GDC approach rather than just GDC-by-GDC modification. My (admittedly) radical proposal for this is as follows.

The first requirement of any generalized design criteria would seem to be to have high-level overall design objectives. For the GDC to be risk-informed, these should be risk-acceptance criterial that are to be achieved by the overall design. As currently configured, the design objectives are to meet various figures-of-merit within the DBAs which are not strictly risk criteria.

The first item of my proposal, then, is to properly define regulatory risk acceptance criteria related to the entire range of regulatory objectives:

- prompt fatalities,
- latent fatalities,
- total deaths,
- injuries,
- land contamination, and (perhaps)
- frequency of release of all magnitudes of radioactivity (e.g. f-c curves). [Note that such f-c curve criterion could encompass all of the above as well as normal operation releases].

In the specification of these risk-acceptance criteria, it is important to properly account for uncertainty in their determination. The most straight-forward way to accomplish this is to express them in terms of confidence levels at which the risk metrics are to be achieved.

The second part of my proposal, then, is to replace the DBAs with a requirement to have a PRA that evaluates the entire spectrum of accident sequences and develops, to the extent possible, the uncertainty associated with the risk results.

It may come as a surprise to some that my proposal does not include the classical defense-in-depth concept of specifying an acceptance value on core damage frequency (CDF). How, then, is it to be assured that there is an appropriate balance between CDF and conditional containment failure probability (CCFP)? I maintain that, in most reactor concepts, the requirement for the designer to achieve a very high confidence level in the overall risk-acceptance criteria will result in the designer putting most emphasis on the least uncertain parts of the risk determination. Under the current knowledge base and PRA state-of-the-art for LWRs, that would mean much more emphasis on achieving a low value of CDF. The balance would automatically be achieved in order to meet the high confidence levels at a reasonable margin for the risk acceptance criteria as well as because it is in the best economic and operational interests of the licensee to have a low CDF. Note, however, that this is not equivalent to specifying a separate CDF acceptance criterion. The designer may choose (based on operational or economic considerations) to design his system without a containment so long as he meets the various risk-acceptance criteria at the specified confidence levels. Thus, with this concept, the traditional measure of defense-in-depth as a balance between CDF and CCFP is abandoned. Putting the emphasis in design areas that minimize the uncertainty in the risk determination is a sort of defense-in-depth concept.

I would define defense-in-depth as the design philosophy used to assure acceptable uncertainty in the determination of the extent to which risk acceptance criteria are met. This generally consists of providing highly reliable, redundant, and diverse multiple ways to accomplish certain functions important to safety, such as:

- shutdown of reactor power
- removal of decay heat (long and short term)
- control of fission product release.

Classical defense-in-depth would still be invoked by recognizing that there are some events for which full reliance should not be currently placed on the PRA bottom line risk results. These events are those that are of potentially high consequences and for which both the frequency and consequence determination has very high aleatory (knowledge based) uncertainty. For these, the GDC should provide assurance that the frequencies are controlled to acceptable levels. These are events like:

1. Rapid increases in power (power coefficient control).
2. Prolonged chemical reaction driven releases (e.g. $Zr-H_2O$; fuel or moderator combustion; coolant combustion).
3. Significant rapid phase change pressurization (thermal explosions)
4. External events.

The GDC should require that these be controlled to a specified confidence level on acceptable frequency via a DID approach that includes high reliability of SSCs, redundancy, and diversity.

In addition, there are regulatory needs related to design that are not well treated in PRAs. These include things like QA, monitoring, inspectability, and testing. To deal with these in a risk-informed manner will require a risk-based determination of those SSCs that are important to safety. The use of RAW and F-V importance measures are appropriate for this determination but may need augmentation by other importance measures (such as DIM) and by judgment of an expert panel. For these SSCs, the appropriate specification of requirements for QA, etc. would stay very much like they are in the current deterministic GDC.

In summary, then, my concept of risk-informing the GDC consists of the following:

1. Specification of risk-acceptance criteria in terms of confidence levels - recognizing that the choice of confidence levels should accommodate the lack of ability to fully determine both epistemic and aleatory uncertainty.

2. Assessment of the achievement of the above by full scope PRA with uncertainty analysis.

3. Recognition that the risk determinations of some events have too much aleatory uncertainty to rely solely on the above PRA bottom lines. Specification that the frequencies of these be controlled to an acceptable confidence level via the use of reliability, redundancy, and diversity.

4. Recognition that some regulatory objectives are not well suited for PRA (e.g. QA, monitoring, inspection, and testing). Retention of the deterministic requirements for these on SSCs important to safety as determined through appropriate importance measures.

As a perhaps superficial example, lets see how these concepts might apply to the licensing of a gas-cooled pebble bed, modular reactor (PBMR).

I. With respect to Item 1, above, the risk acceptance criteria established in general for a single plant of any kind must be lowered to account for multiple modules by dividing each by the number of modules.

II. With respect to Item 3:

- The strong negative power coefficient driven by a strong temperature coefficient virtually prohibits rapid power increases and oscillations.

- Particular attention must be paid to the protection and control systems because redundancy and diversity are difficult with this design.

- The choice of coolant and fuel eliminates any potential for phase change (thermal) explosions.

- The strong temperature coefficient and thermal capacity assures that the power is turned off prior to fuel damage and fission product release (however, must assure the proper fuel quality is achieved and that the level of contamination of stray uranium is sufficiently low).

- There is a potential for air-ingression accidents to create both a fuel and moderator fire. Much attention must be placed on rendering these events to a low frequency (e.g. of the order of 10^{-6}/yr at 95% confidence level). Particular attention must be paid to external events here, and there must be provision (redundant and diverse) to quench any fire.

- The issue of whether or not a PBMR must have a containment depends on the level of confidence to be attached to meeting the various risk acceptance criteria. If all of these are met at the specified high level of confidence, then a containment is not necessary for licensing.

This simple example shows that this concept of GDC appears to achieve the desired objectives of
- giving the designer appropriate guidance,
- providing a regulatory handle to deal with the substantial issues,
- providing a technically sound way to deal with the issue of need for a containment, and
- providing an appropriate level of defense-in-depth.

APPENDIX B TO PART 50 -- QUALITY ASSURANCE CRITERIA FOR NUCLEAR POWER PLANTS AND FUEL REPROCESSING PLANTS

Introduction. Every applicant for a construction permit is required by the provisions of §50.34 to include in its preliminary safety analysis report a description of the quality assurance program to be applied to the design, fabrication, construction, and testing of the structures, systems, and components of the facility. Every applicant for an operating license is required to include, in its final safety analysis report, information pertaining to the managerial and administrative controls to be used to assure safe operation. Nuclear power plants and fuel reprocessing plants include structures, systems, and components that prevent or mitigate the consequences of postulated accidents that could cause undue risk to the health and safety of the public. This appendix establishes quality assurance requirements for the design, construction, and operation of those structures, systems, and components. The pertinent requirements of this appendix apply to all activities affecting the safety-related functions of those structures, systems, and components; these activities include designing, purchasing, fabricating, handling, shipping, storing, cleaning, erecting, installing, inspecting, testing, operating, maintaining, repairing, refueling, and modifying.

As used in this appendix, "quality assurance" comprises all those planned and systematic actions necessary to provide adequate confidence that a structure, system, or component will perform satisfactorily in service. Quality assurance includes quality control, which comprises those quality assurance actions related to the physical characteristics of a material, structure, component, or system which provide a means to control the quality of the material, structure, component, or system to predetermined requirements.

I. Organization

The applicant (Note 1) shall be responsible for the establishment and execution of the quality assurance program. The applicant may delegate to others, such as contractors, agents, or consultants, the work of establishing and executing the quality assurance program, or any part thereof, but shall retain responsibility therefor. The authority and duties of persons and organizations performing activities affecting the safety-related functions of structures, systems, and components shall be clearly established and delineated in writing. These activities include both the performing functions of attaining quality objectives and the quality assurance functions. The quality assurance functions are those of (a) assuring that an appropriate quality assurance program is established and effectively executed and (b) verifying, such as by checking, auditing, and inspection, that activities affecting the safety-related functions have been correctly performed. The persons and organizations performing quality assurance functions shall have sufficient authority and organizational freedom to identify quality problems; to initiate, recommend, or provide solutions; and to verify implementation of solutions. Such persons and organizations performing quality assurance functions shall report to a management level such that this required authority and organizational freedom, including sufficient independence from cost and schedule when opposed to safety considerations, are provided. Because of the many variables involved, such as the number of personnel, the type of activity being performed, and the location or locations where activities are performed, the organizational structure for executing the quality assurance program may take various forms provided that the persons and organizations assigned the quality assurance functions have this required authority and organizational freedom. Irrespective of the organizational structure, the individual(s) assigned the responsibility for assuring effective execution of any portion of the quality assurance program at any location where activities subject to this appendix are being performed shall have direct access to such levels of management as may be necessary to perform this function.

II. Quality Assurance Program

The applicant shall establish at the earliest practicable time, consistent with the schedule for accomplishing the activities, a quality assurance program which complies with the requirements of this appendix. This program shall be documented by written policies, procedures, or instructions and shall be carried out throughout plant life in accordance with those policies, procedures, or instructions. The applicant shall identify the structures, systems, and components to be covered by the quality assurance program and the major organizations participating in the program, together with the designated functions of these organizations. The quality assurance program shall provide control over activities affecting the quality of the identified structures, systems, and components, to an extent consistent with their importance to safety. Activities affecting quality shall be accomplished under suitably controlled conditions. Controlled conditions include the use of appropriate equipment; suitable environmental conditions for accomplishing the activity, such as adequate cleanness; and assurance that all prerequisites for the given activity have been satisfied. The program shall take into account the need for special controls, processes, test equipment, tools, and skills to attain the required quality, and the need for verification of quality by inspection and test. The program shall provide for indoctrination and training of personnel performing activities affecting quality as necessary to assure that suitable proficiency is achieved and maintained. The applicant shall regularly review the status and adequacy of the quality assurance program. Management of other organizations participating in the quality assurance program shall regularly review the status and adequacy of that part of the quality assurance program which they are executing.

III. Design Control

Measures shall be established to assure that applicable regulatory requirements and the design basis, as defined in §50.2 and as specified in the license application, for those structures, systems, and components to which this appendix applies are correctly translated into specifications, drawings, procedures, and instructions. These measures shall include provisions to assure that appropriate quality standards are specified and included in design documents and that deviations from such standards are controlled. Measures shall also be established for the selection and review for suitability of application of materials, parts, equipment, and processes that are essential to the safety-related functions of the structures, systems and components.

Measures shall be established for the identification and control of design interfaces and for coordination among participating design organizations. These measures shall include the establishment of procedures among participating design organizations for the review, approval, release, distribution, and revision of documents involving design interfaces.

The design control measures shall provide for verifying or checking the adequacy of design, such as by the performance of design reviews, by the use of alternate or simplified calculational methods, or by the performance of a suitable testing program. The verifying or checking process shall be performed by individuals or groups other than those who performed the original design, but who may be from the same organization. Where a test program is used to verify the adequacy of a specific design feature in lieu of other verifying or checking processes, it shall include suitable qualifications testing of a prototype unit under the most adverse design conditions. Design control measures shall be applied to items such as the following: reactor physics, stress, thermal, hydraulic, and accident analyses; compatibility of materials; accessibility for inservice inspection, maintenance, and repair; and delineation of acceptance criteria for inspections and tests.

Design changes, including field changes, shall be subject to design control measures commensurate with those applied to the original design and be approved by the organization that performed the original design unless the applicant designates another responsible organization.

IV. Procurement Document Control

Measures shall be established to assure that applicable regulatory requirements, design bases, and other requirements which are necessary to assure adequate quality are suitably included or referenced in the documents for procurement of material, equipment, and services, whether purchased by the applicant or by its contractors or subcontractors. To the extent necessary, procurement documents shall require contractors or subcontractors to provide a quality assurance program consistent with the pertinent provisions of this appendix.

V. Instructions, Procedures, and Drawings

Activities affecting quality shall be prescribed by documented instructions, procedures, or drawings, of a type appropriate to the circumstances and shall be accomplished in accordance with these instructions, procedures, or drawings. Instructions, procedures, or drawings shall include appropriate quantitative or qualitative acceptance criteria for determining that important activities have been satisfactorily accomplished.

VI. Document Control

Measures shall be established to control the issuance of documents, such as instructions, procedures, and drawings, including changes thereto, which prescribe all activities affecting quality. These measures shall assure that documents, including changes, are reviewed for adequacy and approved for release by authorized personnel and are distributed to and used at the location where the prescribed activity is performed. Changes to documents shall be reviewed and approved by the same organizations that performed the original review and approval unless the applicant designates another responsible organization.

VII. Control of Purchased Material, Equipment, and Services

Measures shall be established to assure that purchased material, equipment, and services, whether purchased directly or through contractors and subcontractors, conform to the procurement documents. These measures shall include provisions, as appropriate, for source evaluation and selection, objective evidence of quality furnished by the contractor or subcontractor, inspection at the contractor or subcontractor source, and examination of products upon delivery. Documentary evidence that material and equipment conform to the procurement requirements shall be available at the nuclear powerplant or fuel reprocessing plant site prior to installation or use of such material and equipment. This documentary evidence shall be retained at the nuclear powerplant or fuel reprocessing plant site and shall be sufficient to identify the specific requirements, such as codes, standards, or specifications, met by the purchased material and equipment. The effectiveness of the control of quality by contractors and subcontractors shall be assessed by the applicant or designee at intervals consistent with the importance, complexity, and quantity of the product or services.

VIII. Identification and Control of Materials, Parts, and Components

Measures shall be established for the identification and control of materials, parts, and components, including partially fabricated assemblies. These measures shall assure that identification of the item is maintained by heat number, part number, serial number, or other appropriate means, either on the item or on records traceable to the item, as required throughout fabrication, erection, installation, and use of the item. These identification and control measures shall be designed to prevent the use of incorrect or defective material, parts, and components.

IX. Control of Special Processes

Measures shall be established to assure that special processes, including welding, heat treating, and nondestructive testing, are controlled and accomplished by qualified personnel using qualified procedures in accordance with applicable codes, standards, specifications, criteria, and other special requirements.

X. Inspection

A program for inspection of activities affecting quality shall be established and executed by or for the organization performing the activity to verify conformance with the documented instructions, procedures, and drawings for accomplishing the activity. Such inspection shall be performed by individuals other than those who performed the activity being inspected. Examinations, measurements, or tests of material or products processed shall be performed for each work operation where necessary to assure quality. If inspection of processed material or products is impossible or disadvantageous, indirect control by monitoring processing methods, equipment, and personnel shall be provided. Both inspection and process monitoring shall be provided when control is inadequate without both. If mandatory inspection hold points, which require witnessing or inspecting by the applicant's designated representative and beyond which work shall not proceed without the consent of its designated representative are required, the specific hold points shall be indicated in appropriate documents.

XI. Test Control

A test program shall be established to assure that all testing required to demonstrate that structures, systems, and components will perform satisfactorily in service is identified and performed in accordance with written test procedures which incorporate the requirements and acceptance limits contained in applicable design documents. The test program shall include, as appropriate, proof tests prior to installation, preoperational tests, and operational tests during nuclear power plant or fuel reprocessing plant operation, of structures, systems, and components. Test procedures shall include provisions for assuring that all prerequisites for the given test have been met, that adequate test instrumentation is available and used, and that the test is performed under suitable environmental conditions. Test results shall be documented and evaluated to assure that test requirements have been satisfied.

XII. Control of Measuring and Test Equipment

Measures shall be established to assure that tools, gages, instruments, and other measuring and testing devices used in activities affecting quality are properly controlled, calibrated, and adjusted at specified periods to maintain accuracy within necessary limits.

XIII. Handling, Storage and Shipping

Measures shall be established to control the handling, storage, shipping, cleaning and preservation of material and equipment in accordance with work and inspection instructions to prevent damage or deterioration. When necessary for particular products, special protective environments, such as inert gas atmosphere, specific moisture content levels, and temperature levels, shall be specified and provided.

XIV. Inspection, Test, and Operating Status

Measures shall be established to indicate, by the use of markings such as stamps, tags, labels, routing cards, or other suitable means, the status of inspections and tests performed upon individual items of the nuclear power plant or fuel reprocessing plant. These measures shall provide for the identification of items which have satisfactorily passed required inspections and tests, where necessary to preclude inadvertent bypassing of such inspections and tests. Measures shall also be established for indicating the operating status of structures, systems, and components of the nuclear power plant or fuel reprocessing plant, such as by tagging valves and switches, to prevent inadvertent operation.

XV. Nonconforming Materials, Parts, or Components

Measures shall be established to control materials, parts, or components which do not conform to requirements in order to prevent their inadvertent use or installation. These measures shall include, as appropriate, procedures for identification, documentation, segregation, disposition, and notification to affected organizations. Nonconforming items shall be reviewed and accepted, rejected, repaired or reworked in accordance with documented procedures.

XVI. Corrective Action

Measures shall be established to assure that conditions adverse to quality, such as failures, malfunctions, deficiencies, deviations, defective material and equipment, and nonconformances are promptly identified and corrected. In the case of significant conditions adverse to quality, the measures shall assure that the cause of the condition is determined and corrective action taken to preclude repetition. The identification of the significant condition adverse to quality, the cause of the condition, and the corrective action taken shall be documented and reported to appropriate levels of management.

XVII. Quality Assurance Records

Sufficient records shall be maintained to furnish evidence of activities affecting quality. The records shall include at least the following: Operating logs and the results of reviews, inspections, tests, audits, monitoring of work performance, and materials analyses. The records shall also include closely-related data such as qualifications of personnel, procedures, and equipment. Inspection and test records shall, as a minimum, identify the inspector or data recorder, the type of observation, the results, the acceptability, and the action taken in connection with any deficiencies noted. Records shall be identifiable and retrievable. Consistent with applicable regulatory requirements, the applicant shall establish requirements concerning record retention, such as duration, location, and assigned responsibility.

XVIII. Audits

A comprehensive system of planned and periodic audits shall be carried out to verify compliance with all aspects of the quality assurance program and to determine the effectiveness of the program. The audits shall be performed in accordance with the written procedures or check lists by appropriately trained personnel not having direct responsibilities in the areas being audited. Audit results shall be documented and reviewed by management having responsibility in the area audited. Followup action, including reaudit of deficient areas, shall be taken where indicated.

[35 FR 10499, June 27, 1970, as amended at 36 FR 18301, Sept. 11, 1971; 40 FR 3210D, Jan. 20, 1975]

Note 1: While the term "applicant" is used in these criteria, the requirements are, of course, applicable after such a person has received a license to construct and operate a nuclear power plant or a fuel reprocessing plant. These criteria will also be used for guidance in evaluating the adequacy of quality assurance programs in use by holders of construction permits and operating licenses.

NRC FORM 335 (2-89) NRCM 1102, 3201, 3202	U.S. NUCLEAR REGULATORY COMMISSION **BIBLIOGRAPHIC DATA SHEET** *(See instructions on the reverse)*	1. REPORT NUMBER (Assigned by NRC, Add Vol., Supp., Rev., and Addendum Numbers, if any.) NUREG-1755

2. TITLE AND SUBTITLE

SOME OBSERVATIONS ON RISK-INFORMING
APPENDICES A AND B TO 10 CFR PART 50

Prepared for the Advisory Committee on Reactor Safeguards

3.	DATE REPORT PUBLISHED	
	MONTH	YEAR
	January	2002

4. FIN OR GRANT NUMBER

5. AUTHOR(S)

J. N. Sorensen
Senior Fellow

6. TYPE OF REPORT

Technical

7. PERIOD COVERED *(Inclusive Dates)*

8. PERFORMING ORGANIZATION - NAME AND ADDRESS *(If NRC, provide Division, Office or Region, U.S. Nuclear Regulatory Commission, and mailing address; if contractor, provide name and mailing address.)*

Advisory Committee on Reactor Safeguards

U. S. Nuclear Regulatory Commission

Washington, DC 20555-0001

9. SPONSORING ORGANIZATION - NAME AND ADDRESS *(If NRC, type "Same as above"; if contractor, provide NRC Division, Office or Region, U.S. Nuclear Regulatory Commission, and mailing address.)*

Same as above

10. SUPPLEMENTARY NOTES

11. ABSTRACT *(200 words or less)*

This report was prepared for the Advisory Committee on Reactor Safeguards to provide a basis for discussing possible changes to the general design criteria (GDC) of Appendix A to Title 10, Part 50, of the Code of Federal Regulations in order to make them more consistent with a risk- informed regulatory structure. The three broad options identified include (1) changing the scope of the GDC from "important to safety" to "important to risk," (2) modifying individual criteria to address risk more directly, and (3) replacing the GDC with safety goals and risk acceptance criteria. Specific changes to improve the risk focus of individual criteria might involve replacing requirements for redundancy, diversity, and independence with an overall reliability goal. Applicability of the GDC to non-light-water reactors (LWRs) is briefly examined, with the conclusion that slightly more than half of the criteria could apply to non-LWRs, but the remainder should be modified or replaced to address phenomena important to the safety of other reactor types. Another brief discussion of the quality assurance requirements of Appendix B to 10 CFR Part 50 concludes that Appendix B has sufficient flexibility to permit less stringent requirements for SSCs that are less important to risk.

12. KEY WORDS/DESCRIPTORS *(List words or phrases that will assist researchers in locating the report.)*

10 CFR Part 50
General design criteria
GDC
Risk-informed regulation
Risk-informing

13. AVAILABILITY STATEMENT

unlimited

14. SECURITY CLASSIFICATION

(This Page)

unclassified

(This Report)

unclassified

15. NUMBER OF PAGES

16. PRICE

This form was electronically produced by Elite Federal Forms, Inc.

NUREG-1755

SOME OBSERVATIONS ON RISK-INFORMING APPENDICES A AND B TO 10 CFR PART 50

JANUARY 2002

www.ingramcontent.com/pod-product-compliance
Lightning Source LLC
Chambersburg PA
CBHW081841170526
45167CB00007B/2868